Advanced Wireless Communications and Mobile Networks - Current Status and Future Directions

Edited by Naser Ojaroudi Parchin

Published in London, United Kingdom

Advanced Wireless Communications and Mobile Networks - Current Status and Future Directions
http://dx.doi.org/10.5772/intechopen.1006224
Edited by Naser Ojaroudi Parchin

Contributors
Bhargav Appasani, Chan Hwang See, Haleh Jahanbakhsh Basherlou, Leni Joaquim de Matos, Martins Odiamenhi, Ming Shen, Mthulisi Velempini, Naser Ojaroudi Parchin, Pattepu Sunil, Pedro Vladimir Gonzalez Castellanos, Prasanta Kumar Patra, Rodrigo Amitrano Bilobran, Sindiso M. Nleya, Srinivas Ramavath, Tatenda T. Gotora, Umesh Chandra Samal, Vanessa Przybylski Ribeiro Magri, Vítor Luiz Gomes Mota

First published in London, United Kingdom, 2025 by IntechOpen
IntechOpen is the global imprint of INTECHOPEN LIMITED, registered in England and Wales, registration number: 11086078, 167-169 Great Portland Street, London, W1W 5PF, United Kingdom

For EU product safety concerns: IN TECH d.o.o., Prolaz Marije Krucifikse Kozulić 3, 51000 Rijeka, Croatia, info@intechopen.com or visit our website at intechopen.com.

British Library Cataloguing-in-Publication Data
A catalogue record for this book is available from the British Library

Advanced Wireless Communications and Mobile Networks - Current Status and Future Directions
Edited by Naser Ojaroudi Parchin
p. cm.
Print ISBN 978-1-83634-755-2
Online ISBN 978-1-83634-754-5
eBook (PDF) ISBN 978-1-83634-756-9

If disposing of this product, please recycle the paper responsibly.

Meet the editor

Dr. Naser Ojaroudi Parchin is an Assistant Professor at Edinburgh Napier University in the UK. He earned his Ph.D. in Electrical Engineering from the University of Bradford, where he later served as a Postdoctoral Research Assistant and research fellow in the ESA-funded SATNEX V project. From 2018 to 2020, he was a Marie Curie Research Fellow in the H2020-ITN-SECRET project. He is an active member of IEEE, MCAA, EurAAP, HEA, and EPC and reviews for leading journals and publishers. Dr. Naser has authored over 400 journal and conference papers, as well as several books and chapters, with more than 9,000 citations and a h-index of 58. He has received multiple research awards and grants for his contributions to the field.

Contents

Preface **IX**

Section 1
Foundations and Evolution of Wireless Networks **1**

Chapter 1 **3**
Introductory Chapter: Advanced Wireless Communications and Mobile Networks – Current Status and Future Directions
by Naser Ojaroudi Parchin

Chapter 2 **7**
Beyond 5G: The Evolution of Wireless Networks and Their Impact on Society
by Sindiso M. Nleya, Mthulisi Velempini and Tatenda T. Gotora

Section 2
Technologies and Applications in 5G and Beyond **33**

Chapter 3 **35**
5G and Beyond: Advancements in Wireless Communications for IoT and Smart Cities
by Srinivas Ramavath, Umesh Chandra Samal, Prasanta Kumar Patra, Pattepu Sunil and Bhargav Appasani

Section 3
Antenna Design and Measurement for 5G/6G Systems **55**

Chapter 4 **57**
Automatized Measurement Setup to Determine 5G and 6G Radiation Pattern for Miniaturized Arrays Antennas
by Vanessa Przybylski Ribeiro Magri, Leni Joaquim de Matos, Pedro Vladimir Gonzalez Castellanos, Vítor Luiz Gomes Mota and Rodrigo Amitrano Bilobran

Chapter 5 **77**
An Overview of THz Antenna Design for 5G/6G Wireless Communications
by Haleh Jahanbakhsh Basherlou, Martins Odiamenhi, Naser Ojaroudi Parchin, Chan Hwang See and Ming Shen

Preface

The landscape of wireless communication has undergone a remarkable transformation over the last few decades, advancing from analog voice systems to highly intelligent, data-driven, and adaptive networks. This edited book presents a timely and comprehensive exploration of the current status and future directions of wireless communications, with a particular focus on fifth-generation (5G), Beyond 5G (B5G), and sixth-generation (6G) technologies. The book gathers contributions from leading researchers and professionals who offer insights into the technological innovations shaping our increasingly connected world.

The volume begins with Chapter 1: *Introductory Chapter: Advanced Wireless Communications and Mobile Networks – Current Status and Future Directions*, which lays the groundwork by examining the rapid progression of wireless technologies and the societal shifts they enable. It introduces the core themes that are further developed in the following chapters. Chapter 2: *Beyond 5G: The Evolution of Wireless Networks and Their Impact on Society* delves into the generational growth of wireless systems, highlighting the technological shifts and their social implications, while exploring anticipated transformations that 6G might bring. Chapter 3: *5G and Beyond: Advancements in Wireless Communications for IoT and Smart Cities* examines how next-generation networks are empowering IoT ecosystems and smart urban environments through reliable, high-speed, and low-latency connectivity. Chapter 4: *Automatized Measurement Setup to Determine 5G and 6G Radiation Pattern for Miniaturized Arrays Antennas* presents a cost-effective and practical approach for characterizing antenna systems vital for the development and testing of future wireless platforms. Chapter 5: *An Overview of THz Antenna Design for 5G/6G Wireless Communications* explores the challenges and breakthroughs in terahertz antenna design, addressing one of the key enabling technologies for ultra-high-speed wireless networks. Together, these chapters present a holistic view of where wireless communication stands today and where it is headed. They address foundational concepts, technical innovations, and practical implementations across diverse domains, from healthcare and smart infrastructure to environmental sensing and autonomous systems. This book is intended for researchers, engineers, practitioners, and students seeking to understand and contribute to the rapidly evolving wireless communication domain. By offering a blend of theory, application, and critical foresight, it serves as both a reference and an inspiration for further exploration in the field.

I would like to express our sincere gratitude to all contributing authors for their expertise and dedication, and to the reviewers for their thoughtful insights that helped improve the quality of this work. Special thanks also go to the editorial support team at IntechOpen for their guidance throughout the publishing process.

Naser Ojaroudi Parchin
School of Computing,
Engineering and the Built Environment,
Edinburgh Napier University,
Edinburgh, United Kingdom

Section 1

Foundations and Evolution of Wireless Networks

Chapter 1

Introductory Chapter: Advanced Wireless Communications and Mobile Networks – Current Status and Future Directions

Naser Ojaroudi Parchin

1. Introduction

Wireless communication has progressed rapidly in recent decades, transitioning from a basic system for analog voice transmission to a complex, intelligent infrastructure that underpins today's digital society. It now serves as a critical enabler across sectors, supporting innovation, driving economic growth, and reshaping interactions between people and technology [1]. As global reliance on digital services continues to grow, wireless networks have become a key foundation for digital transformation in both developed and developing regions. A major contributor to this evolution is the surge in connected devices. While mobile phones and tablets remain widespread, today's network environment also includes wearable technologies, embedded processors, and billions of real-time Internet of Things (IoT) sensors [2]. These devices generate immense volumes of data, posing new challenges for network capacity, reliability, and scalability. Forecasts suggest that mobile data traffic could exceed 300 exabytes per month by 2030, fueled by widespread use of high-definition video, augmented reality, and cloud-based services [3].

2. Applications and limitations of modern wireless communications

In addition to human-driven applications, machine-type communication (MTC) has seen rapid adoption. MTC enables devices to communicate autonomously, without requiring human intervention. It plays a central role in smart manufacturing, healthcare systems, transportation networks, and infrastructure monitoring [4]. However, these use cases place strict demands on wireless systems, especially in terms of latency, reliability, and energy efficiency, which many earlier mobile generations were not designed to meet [5]. The importance of wireless communication extends into a wide range of essential services. In healthcare, it supports remote diagnostics, telemedicine, and continuous patient monitoring. In the education sector, wireless networks provide access to virtual classrooms and collaborative learning platforms. Transportation systems rely on technologies such as vehicle-to-vehicle (V2V) and vehicle-to-infrastructure (V2I) communication to enable autonomous and connected mobility. Wireless sensing systems also enhance precision agriculture by monitoring soil and environmental conditions. In emergency response scenarios, wireless networks often remain operational even when conventional infrastructure fails, offering critical communication links [6].

3. Future directions: From 5G to 6G

The deployment of fifth-generation (5G) networks represents a major step forward in mobile communication technology. Unlike earlier generations that focused primarily on voice and data services, 5G was designed to accommodate a diverse range of applications. These are categorized into three main service types: enhanced mobile broadband (eMBB), ultra-reliable low-latency communication (URLLC), and massive machine-type communication (mMTC) [7]. To fulfill these demands, 5G networks implement several advanced technologies including millimeter-wave (mmWave) spectrum, massive multiple-input multiple-output (MIMO), network slicing, and edge computing [8]. These advancements have enabled significantly faster speeds, reduced latency, and improved capacity for device connections. As a result, applications such as immersive media, industrial automation, and real-time online services have become increasingly viable. Despite these improvements, the rising complexity of applications, particularly those involving real-time interaction, automation, and data-intensive processing, has exposed some limitations of current 5G systems [9].

To address these challenges, the global research community is now exploring the next generation of wireless networks: sixth generation (6G). Although 6G is still in the early stages of research and definition, it is envisioned as a highly adaptive, intelligent communication environment. Its architecture will likely integrate both terrestrial and non-terrestrial platforms, including low-Earth orbit satellites, high-altitude base stations, and drone-supported systems. These components aim to provide broader coverage, greater reliability, and seamless global connectivity. The anticipated performance targets for 6G are ambitious. Networks are expected to support peak data rates up to 1 terabit per second, offer latency below 1 millisecond, and maintain extremely high reliability. In contrast to previous generations, 6G will embed artificial intelligence (AI), machine learning (ML), and sensing capabilities directly into its core architecture [10]. This integration will enable real-time adaptation to user behavior, environmental context, and service requirements. It is also expected to support advanced use cases such as immersive extended reality (XR), digital twins, and autonomous intelligent systems, pushing the boundaries of what wireless communication can deliver.

Author details

Naser Ojaroudi Parchin
School of Computing, Engineering and the Built Environment, Edinburgh Napier University, Edinburgh, UK

*Address all correspondence to: n.ojaroudiparchin@napier.ac.uk

References

[1] Ericsson. Ericsson Mobility Report. Stockholm, Sweden: Ericsson; 2024. Available from: www.ericsson.com

[2] Gupta A, Jha RK. A survey of 5G network: Architecture and emerging technologies. IEEE Access. 2015;**3**:1206-1232

[3] Andrews JG et al. What will 5G be? IEEE Journal on Selected Areas in Communications. 2014;**32**(6):1065-1082

[4] Giordani M, Polese M, Mezzavilla M, Rangan S, Zorzi M. Toward 6G networks: Use cases and technologies. IEEE Communications Magazine. 2020;**58**(3):55-61

[5] Saad W, Bennis M, Chen M. A vision of 6G wireless systems: Applications, trends, technologies, and open research problems. IEEE Network. 2020;**34**(3):134-142

[6] David K, Berndt H. 6G vision and requirements: Is there any need for beyond 5G? IEEE Vehicular Technology Magazine. 2018;**13**(3):72-80

[7] Popovski P et al. Wireless access for ultra-reliable low-latency communication: Principles and building blocks. IEEE Network. 2018;**32**(2):16-23

[8] Rappaport TS et al. Wireless communications and applications above 100 GHz: Opportunities and challenges for 6G and beyond. IEEE Access. 2019;**7**:78729-78757

[9] Agiwal M, Roy A, Saxena N. Next generation 5G wireless networks: A comprehensive survey. IEEE Communications Surveys & Tutorials. 2016;**18**(3):1617-1655

[10] Wang C-X et al. Artificial intelligence enabled wireless networking for 5G and beyond: Recent advances and future challenges. IEEE Wireless Communications. 2020;**27**(1):16-23

Chapter 2

Beyond 5G: The Evolution of Wireless Networks and Their Impact on Society

Sindiso M. Nleya, Mthulisi Velempini and Tatenda T. Gotora

Abstract

Wireless network generations are defined by advancements in transmission technology and the utilization of different frequency bands over time. During the development of numerous wireless network generations, they have significantly impacted society by influencing how people live, learn, work and communicate. In the different phases of wireless network generations, from the first generation (1G) to the current fifth generation (5G), some serious challenges have affected the operation and efficiency of networks. This chapter traces the evolution of wireless networks from their inception to the advent of 5G and beyond. It examines the key technological advancements that have driven this evolution, including the shift from analog to digital, the emergence of mobile broadband, and the increasing reliance on higher frequencies and advanced signal processing techniques. Furthermore, the chapter analyzes the societal and economic impacts of each generation of wireless technology, highlighting the transformative effects on communication, commerce, and daily life. Finally, the chapter delves into the potential of 6G and beyond, discussing emerging technologies like terahertz communication, holographic beamforming, and AI-powered network management, and their potential to revolutionize industries such as healthcare, transportation, and manufacturing. The chapter concludes by discussing the challenges and opportunities associated with this rapid technological advancement.

Keywords: wireless technology evolution, fifth generation, sixth generation, communication, society

1. Introduction

Ever since the inception of the initial radio communication system broadcast originating from a provisional radio installation Marconi established on the Isle of Wight in 1895 to the advanced 5G networks of today, wireless networks have experienced a tremendous evolution. This evolution has spanned multiple generations in approximately 40 years, from first generation (1G) to fifth generation (5G) [1–4], and is currently evolving beyond 5G(B5G) [5] into the sixth generation (6G). At its core, this evolution has brought with it an array of emerging communication technologies that affect individuals, groups, enterprise organizations, and entire societies [6]. Future developments are anticipated to bring higher data speeds, better coverage, cost-effective resource usage, better security, adaptability, and scalability to 5G and B5G wireless networks [7]. The advent of the B5G\6G era marks an unprecedented milestone in wireless communication. As the telecommunications sector advances beyond 5G and begins 6G deployment, researchers and industry pioneers are actively engaged in shaping 6G and the groundbreaking technologies that will define it [8]. 6G is predicted to bring hyper-speed networks, real-time communications, and the integration of technologies such as artificial intelligence (AI), augmented reality (AR), virtual reality (VR), and more [8]. The transformative impact of wireless technologies can have a deep impact on many aspects of life. 6G is not just about building a faster, more efficient network; it is about embedding sustainability into the very core network makeup: how it can drive positive environmental, societal, and economic change [9, 10]. Moreover, [11] highlights the urgent need for 6G networks to address the burgeoning demands of cutting-edge applications like the Internet of Things (IoT), virtual reality, and ultra-high-definition video. While the paper presents a compelling vision for 6G, it suffers from a lack of critical evaluation and concrete proof to back up the viability and practical implementation of the proposed goals.

This chapter explores empowering technologies such as Massive Multiple Input Multiple Output (MIMO), Terahertz (THz) communications, Edge computing, and Artificial Intelligence (AI), which are undoubtedly crucial for 6G development. However, the discussion of these technologies remains largely superficial. It lacks a deeper dive into their potential drawbacks, application challenges, and the trade-offs associated with their implementation [12]. This chapter is organized as follows: Section 2 traces the evolution of wireless technology, starting with first-generation systems and progressing through to the current fifth generation and beyond. Section 3 explores the next frontier beyond 5G, outlining the key advancements and technologies expected to shape future wireless communication. Section 4 examines the societal impact of wireless network advancements, highlighting the transformative effects on various aspects of human life. Section 5 delves into the challenges and considerations associated with the evolving wireless landscape, such as spectrum allocation, security, and energy efficiency. Section 6 concludes the chapter with a summary of key findings and insights.

2. The evolution of wireless network

The evolution of wireless cellular technology, known as Gs, has progressed approximately every decade. 1G emerged before 1990, followed by 2G in 1990, 3G in 2000, 4G in 2010, and 5G in 2020, with the anticipation of beyond 5G technologies by 2030 [13]. The evolution of wireless technology has been a remarkable journey, characterized by several key aspects, including:

Variable bandwidth and speed: depicted by a transition from analog to digital signals resulting in increased transmission rates and subsequent generational advancements. Each generation (1G to 5G) has brought substantial increases in data speeds, enabling more demanding applications like video streaming, high-definition video calls, and augmented/virtual reality [14, 15]. Furthermore, the generational advancements have been in tandem with frequency increases with the utilization of higher frequency bands (e.g., mmWave) in 5G and beyond, promising increasingly faster speeds and lower latency [16].

- *Enhanced coverage and capacity*: This aspect has been represented by network expansion, cell tower densification and massive MIMO. Network expansion has been seen in the form of continuous expansion of network coverage to reach more users in more locations while cell tower densification has resulted from increased deployment of cell towers to improve coverage and capacity [17]. The massive MIMO employs multiple antennas at both the transmitter and receiver to increase capacity and improve coverage [18].

- *Reduced latency*: Real-time applications: Lower latency is crucial for real-time applications like autonomous vehicles, remote surgery, and industrial automation. Advanced Techniques: Techniques like network slicing and edge computing are being employed to minimize latency.

- *Increased connectivity*: Connectivity has evolved subsequently leading to increased connectivity, especially in the case of 5G [19]. This has been manifested by increased IoT growth [20, 21]. Machine-to-Machine Communication [22] and ability to support diverse resource-hungry applications [23].

- *Enhanced security*: Applying cryptography in the network will result in strong encryption and authentication implementations that protect user data and network integrity [24, 25]. Through utilizing AI and machine learning for threat detection and mitigation will also improve network security [26].

- *Integration with other technologies*: The evolution of performance needs across generations drives the development of innovative network architectures that seamlessly integrate current and future radio access technologies [27, 28].

- *Focus on user experience*: Improved Quality of Service (QoS) by provisioning consistent and reliable service to all users [29, 30]. Furthermore, personalized experiences are tailored to meet individual user needs and preferences by adapting the network performance to each user scenario.

2.1 First generation (1G)

The evolutionary journey of wireless communication networks was initially recorded in the 1970s with the first generation of cellular networks [31]. Fundamentally, the first generation did fulfill the basic mobile voice calls. The infrastructural network comprised of analog base stations utilizing Frequency Division Multiple Access (FDMA) for signal transmission and reception. Narrowband analog frequencies were used to transmit the voice messages utilizing these base stations distributed throughout the service area [32]. Practically, these networks were initially deployed in Japan by Nippon Telephone and Telegraph Company (NTT) in Tokyo in 1979 and subsequently spread to the US, Finland, the UK, and Europe [33]. Using analog signals, this generation had the following features [34]:

- Frequencies: 800 and 900 MHz.

- Bandwidth of 10 MHz, featuring 666 duplex channels with a 30 KHz bandwidth each.

- Technology: Analog switching.
- Modulation: FM (Frequency Modulation).
- Service mode: voice-only [34].

The popular systems of 1G included the European Total Access Communication System (ETACS), Nordic Mobile Phone System (NMTS), Advanced Mobile Phone System (AMPS), and Total Access Communication System (TACS) [34]. These 1G systems were, however, constrained by poor voice quality owing to interference as well as poor battery life. Moreover, the large-sized mobile phones were not convenient to carry and had less security. The network had a limited number of users and cell coverage with roaming between similar systems is almost impossible.

2.2 Second generation (2G)

A fundamental part of this evolutionary stage is the introduction of a new digital technology for wireless transmission called the Global System for Mobile Communication (GSM). The GSM standard enabled a maximum data rate of 14.4 to 64 kbps, which is adequate for email and short messaging services (SMS) [34]. The 2G system's salient characteristics are:

- Digital switching system.
- SMS services are supported.
- Roaming capabilities are available.
- Improved security features.
- Encrypted voice communication.
- Initial internet services with low data rates.

Drawbacks of the 2G systems included:

- Low data rate.
- Restricted mobility.
- Restricted features on mobile devices.
- Limited hardware capacity and few users.

The popular systems of the 2G were aimed at improving data rates ranging from General Packet Radio Service (GPRS), Enhanced Data GSM Evolution (EDGE) and CDMA2000 [34, 35].

2.3 Third generation (3G)

A 3G network refers to the third generation of mobile network technology that utilizes Universal Mobile Telecommunications System (UMTS) as its core network architecture [36]. Building upon their 2G predecessors, 3G networks introduced markedly higher data rates and enhanced bandwidth, catering to the growing demands of the expanding smartphone market. Additionally, in the realm of IoT, 3G-enabled data-heavy applications like video transmission [32, 37]. Rajiv [34] describes 3G popular systems as High-Speed Downlink Packet Access (HSDPA) and High-Speed Uplink Packet Access (HSUPA) with key features as:

- Increased data rate.
- Video calling.
- Enhanced security, more users, and coverage.
- Mobile application support.
- Multimedia message support.
- Location tracking and maps.
- Better web browsing.
- TV streaming.
- High-quality 3D games.

2.4 Fourth generation (4G)

4G systems are improved versions of IEEE-developed 3G networks, they offer increased data rates and are capable of handling more sophisticated media services. Long-Term Evolution (LTE) and LTE advanced wireless technologies are used in 4G systems. Major highlights of 4G services relevant to mobile users are [38]:

- Flexible application adaptability and dynamic user traffic.
- Enhanced spectrum environment.
- Support for multiple network protocols.
- Efficient radio access interfaces.
- Improved service quality.
- Facilitation of interactive media, voice and video streaming, gaming, internet access, and various broadband services.
- IP-enabled mobile system which exhibits high capacity and low cost per bit.
- Global access.
- Better service portability.
- Improved congestion avoidance mechanisms.

Technology Feature	1G	2G	3G	4G	5G	Beyond 5G
Start/ Deployment	1970–1980	1990–2004	2004–2010	2010–2019	2020	Expected around 2030
Data bandwidth	2 kbps	64 kbps	2 Mbps	1 Gbps	10 Gbps	100 Gbps
Technology	Analog Cellular	Digital Cellular Technology	CDMA 200	Wi-max, LTE, Wi-Fi	Next-Generation Cellular (NR)	Hypothetical, under research and development
Service	Voice Mobile telephony	Enhanced digital voice, mobile services, and higher capacity for packetized data.	Seamlessly integrated premium audio, video, and data	Access to information and wearable technology.	Advanced information access and AI-powered wearable devices.	wide range of advanced services
Multiplexing	FDMA	TDMA, CDMA	CDMA	CDMA	CDMA	CDMA
Switching	Circuit	Circuit, Packet	Packet	All Packet	All Packet	Advanced switching technologies
Core network	PSTN	PSTN	Packet	Internet	Internet	AI-driven

Table 1.
Wireless Technology Evolution Comparison.

4G's all-IP architecture guarantees compatibility with standard networks, integrates support for 3G and analog wireless technologies and requires reasonable Capital expenditure (CAPEX) to integrate with the 4G systems.

2.5 Fifth generation (5G)

5G, the fifth mobile network introduced since 2020, is rapidly being adopted and implemented by numerous nations worldwide as a new technological development. Strong internet connection without lag or disruption is possible with a 5G network. Furthermore, 5G will revolutionize society through its application in self-driving cars and virtual reality [2]. 5G is among the most advanced technologies that would transform communication between humans, machines, and sensors to realize a connected, smarter, and safer world.

The 5G technology's essential features include:

- Blazing-fast mobile internet speeds reaching up to 10 Gbps.
- Low latency in milliseconds is essential for mission-critical applications.
- Reduced data costs.
- Enhanced security and dependable network.
- Employs small cells and beamforming technologies to enhance efficiency.
- Forward-compatible networks enable future advancements and upgrades.
- Cloud-based infrastructure provides energy efficiency, effortless maintenance, and seamless hardware upgrades [34].

A comprehensive contrast between the generations is delineated in **Table 1.**

3. Beyond 5G

Beyond 5G, is what could be called the sixth generation (6G), this generation is envisioned to have applications and key technologies that will enable seamless interaction between humans and the cyber-physical world. NEC [38] defines B5G as a communication system that integrates networks with distributed data processing, organically utilizing computation resources distributed worldwide and implementing real-time interaction across the globe. Beyond 5G (B5G) communication networks will fundamentally reshape society, eliminating physical communication barriers and unlocking human potential on an unprecedented scale. This technical evolution signifies a transformative leap, far exceeding a mere advancement in wireless technology.

B5G will evolve by integrating cutting-edge network solutions with complementary technologies like distributed computing and artificial intelligence (AI). This advancement in technical capabilities promises profound impacts, ranging from improving individual lives to revolutionizing businesses and reshaping society on a broader scale [38]. B5G will be supported by three types of communication: hyper-realistic communication, digital twins, and ubiquitous global coverage.

Hyper-connectivity refers to the interconnected nature of digital environments, encompassing the interaction among information systems, data, and devices, all linked together through the internet [39, 40]. Hyper-connected communications have no constraints on the data rate, coverage and computing. It is expected that B5G networks will realize hyper-connectivity for various highly demanding applications such as immersive augmented reality (AR) and virtual reality (VR), mobile hologram, and metaverse [41–43].

The key candidate technologies for realizing the hyper-connected society include Terahertz (THz) communications, integrated non-terrestrial and terrestrial networks, distributed computing for connected intelligence and advanced multiple access techniques. Digital twins are another form of communication, and they capture real-world data and replicate it within a virtual environment. Ultimately, a variety of technologies, including High-altitude platform stations (HAPS), low earth orbit (LEO) satellites, terahertz waves, and geostationary orbit satellites (GEO), will be used to eradicate any residual dead zones and achieve seamless coverage throughout the entire planet [38].

The key 6G technologies include Artificial Intelligence, Optical Radio Technology, Terahertz communication, Multiple input and Multiple output (MIMO) technique, free-space optics (FSO) backhaul technique, High visibility 3D connecting and Blockchain [41–44].

- *Artificial intelligence*: The development of B5G networks depends on AI and ML since they make it possible to automate and optimize network operations. For instance, ML algorithms may identify security vulnerabilities before they cause harm, and AI can forecast network traffic patterns and dynamically modify capacity to preserve flawless user experiences [45].

- *Terahertz communication*: Future networks might use terahertz frequencies, even though 5G depends on millimeter waves. With their much greater bandwidth, these waves allow for faster data transfer and support for more devices at once.

- *Optical radio technology*: This technology provides high-density broadband communication services, offering significant advantages including inherent physical layer security, ultra-low latency, immunity to electromagnetic interference, cost-effectiveness, access to extensive unlicensed spectrum, and simplified deployment. The optical band, encompassing visible light, ultraviolet (UV), and infrared (IR), presents diverse applications and considerations [46]. With applications in indoor, underwater, vehicular, and long-distance communications, Optical wireless technologies, including Light Fidelity (Li-Fi), Visible Light Communication (VLC), Optical Camera Communication (OCC), Light Detection and Ranging (LiDAR), and Free-Space Optics (FSO) have become very popular [46]. The electrical bandwidth of optoelectronic devices limits the usage of optical bands, although they have three orders of magnitude more spectrum resources than RF bands. Data speeds above 1 Gbps have been made possible by advancements in high-performance devices, such as silicon photomultipliers and fast organic light-emitting diodes (OLEDs).

- *MIMO technique*: Massive MIMO's amazing increase of spectral efficiency (SE), first coined by Marzetta in 2009 [47], has been a key technology for 5G.

Ultra-massive MIMO offers significant potential with applications such as enhanced multiplexing, improved interference mitigation, increased energy efficiency, and expanded coverage, including non-terrestrial networks. Its unique ability to spatially resolve signals with high precision enables accurate 3D positioning in challenging wireless environments.

4. Societal impact of 5G and future networks

5G and future networks are poised to have a significant societal impact, provided that policy and regulation facilitate their deployment [48].

4.1 Hyper-connected society

The transition from 5G to 6G and beyond represents a giant step toward a hyper-connected world where the lines between the physical and digital blur. Practically, a hyper-connected society involves seamless connectivity across devices and environments. Future advancements could translate to intelligent edge computing wherein processing data closer to where it is generated to reduce latency and bandwidth usage. Simultaneously achieving global coverage with internet provision through satellite technology and other means, ensuring that even the most remote locations are connected [49].

4.2 Enhanced industries

6G is expected to bring a wide range of development and new opportunities across industries [50, 51]. The range of possible uses for 6G technology includes: extremely high-speed internet, IoT and smart cities, autonomous vehicles, virtual and augmented reality, telemedicine and remote healthcare, industrial automation and manufacturing as well as remote sensing and monitoring. Concerning the issue of provisioning of extremely high-speed internet, the speeds are expected to reach 1Tbps facilitating ultra-faster data transfer and streaming. Furthermore, this technology allows efficient and effective communication between devices.

IoT and smart cities are other applications brought about by this technology. This is achieved through supporting and connecting more devices and this drives the development and subsequent development of new IoT applications and services. B5G also plays a significant role in the development of smart cities, wherein a range of devices and systems are interconnected and work in concert to enhance citizens' quality of life. The automotive industry is another case in point wherein vehicles are made fully autonomous. Cars, drones and public transit systems are capable of seamless communication, analyzing live data and subsequently making decisions to improve safety. These systems are also configured to alleviate congestion as well as enhance the travel experience. Further enhancements are also expected in virtual and augmented reality owing to the abundance of bandwidth. VR and AR technologies will thus become more immersive and effective. Elsewhere in telemedicine and remote healthcare, communication between healthcare givers and patients is set to be faster and more effective, rendering remote monitoring and treatment more feasible.

In the manufacturing sector enhancement will be in the form of acceleration of industry 4.0 implementation in manufacturing, encompassing IoT, big data, and AI.

Thus, there is efficient and productive manufacturing due to the improved connectivity. Finally, in remote sensing and monitoring, 6G will enable faster, more reliable, and more accurate remote sensing and monitoring, and this is exemplified by, weather forecasting, natural resource management, and environmental monitoring.

4.3 New experiences

Immersive communications will come to reality and shift the current communication paradigm in three aspects, namely immersive experiences, holographic communication and personalized experiences [52]. Immersive experiences will increasingly dissolve the line between physical and virtual realms, enabling innovative ways to interact across both worlds. Security communication and sensory linkage are implied by immersive experience, which will enable dynamic three-dimensional interaction. Visual representations of people, objects, and their surroundings can be realistic and natural, transcending time and space and incorporating the senses of taste, smell, and touch [53]. Mobile network operators will be able to provide consumers with a customized network experience thanks to the AI-driven radio access network (RAN), which is based on real-time user data gathered from several sources. By personalizing quality-of-experience (QoE) and quality-of-service (QoS), operators can further utilize real-time user data to enhance superior services.

AI can be used by the operators to customize a range of services [54]. The authors in [55] refer to this kind of experience as a seamless digital experience wherein converging the physical and digital worlds, digital sensors will enable digital representations to create digital twins of people, real items, and their environments. B5G will enhance metaverse and mixed reality experiences for users by delivering the necessary speed and efficiency.

4.4 Enhanced disaster management

Unmanned aerial vehicles (UAVs) can be used to aid communication in emergency scenarios when there is a well-designed backhaul network, which means that Unmanned Aerial Base Stations (UABSs) can connect directly to the core network while providing connectivity using frequency bands in the 3.5Ghz and 60Ghz. The UABS might also be used in remote locations or hard-to-reach areas where connectivity is a problem. Mobile network operators (MNOs) can invest in UABS and have a swarm of solar-powered UAVs flying around to increase network capacity and coverage, which will ensure connectivity is maintained even when disasters strike like cyclones and earthquakes occur [56].

4.5 Economic growth

Increasingly, technology is becoming the epicenter of global competition and economic growth. B5G, the next generation of mobile technology, will define the economic landscape for 2030 and beyond [57]. By 2030, the worldwide 6G market is anticipated to reach a value of over $1 trillion. New business models, industrial applications, and the accelerating adoption of new technology will all be necessary for this kind of growth. 6G, which promises even faster speeds, significantly lower latency, and more reliable networks, is seen to have the potential to launch entirely new sectors, just like 4G did with the development of social media, the app economy, and many other things. According to estimates from the United Nations Development

Sector	Potential Impact	Economic Benefits
Healthcare	Remote surgeries, AI-driven diagnostics, faster health data analysis.	Enhanced efficiency, cost reductions, improved patient outcomes, increased accessibility to healthcare, advancements in innovation and research, and economic growth.
Education	Virtual and augmented reality-based learning, real-time global classrooms.	Equalized learning opportunities, and enhanced education quality.
Manufacturing	AI-driven smart factories, predictive maintenance, Industry 5.0. [59]	Improved productivity, reduced costs, and minimized downtime.
Transport	Fully autonomous vehicles and smart traffic management systems.	Boosted Efficiency, Reduced Costs, Improved Safety, Innovation and Growth, Employment Opportunities, and Enhanced Global Competitiveness., Reduced accidents, better fuel efficiency, and reduced emissions.
Smart City	Ultra-high-speed connectivity, IoT integration, Advanced AI and Machine Learning Technologies, Immersive Experiences, Sustainable Development and global connectivity	Increased productivity, job creation, innovation and entrepreneurship, cost savings, investment, enhanced public services, and global competitiveness.

Table 2.
Potential impacts of 6G in other sectors.

Program (UNDP) and the International Telecommunication Union (ITU), digital technologies like cybersecurity, artificial intelligence, and digital infrastructure could directly help achieve 70% of the Sustainable Development Goals (SDGs) targets [58]. **Table 2** depicts the potential impacts of 6G in sectors such as healthcare, education, manufacturing, smart city and transportation.

In the healthcare sector, 6G has the potential to provide real-time patient monitoring by leveraging on wearable devices and medical sensors. Health personnel will thus be able to track and monitor patients for any abnormalities to timely intervene. Moreover, the availability of high connectivity and low latency makes telemedicine and remote monitoring possible. Telemedicine and remote monitoring provide healthcare to rural and underserved areas [60]. Additionally, the advancements in 6G will facilitate advanced medical imaging such as high-resolution Magnetic Resonance Imaging (MRI) and Computed Tomography (CT) scans. The increasing developments in AI will make AI-driven healthcare wherein machine learning algorithms can be leveraged upon together with appropriate datasets to provide predictive analytics, personalized treatment and early disease detection. AI will also enable robotic surgery wherein surgeons have the capacity to perform complex procedures remotely with increased accuracy and control. Finally, with the existence of numerous data sources in healthcare, the seamless integration of health data from numerous sources such as wearable devices, electronic health records and medical imaging systems feasible [61]. The transportation sector stands to benefit significantly from 6G technology, including improved connectivity, autonomous vehicles, intelligent traffic management, cooperative collision avoidance, remote vehicle control, and environmental advantages [62]. The improved connectivity will bring about intelligent transportation systems that will enable intelligent traffic management. Furthermore, autonomous vehicles will easily communicate, interact and coordinate within the infrastructure. Similarly, remote vehicle control is enabled, opening

Consideration	Comments
Network Infrastructure and Accessibility	Expanding connectivity in remote or underserved areas involves using a combination of satellite, terrestrial, and aerial networks to enhance coverage.
Affordability and Accessibility of Devices	Affordable and accessible 6G-enabled devices are vital for ensuring widespread adoption across all socioeconomic groups.
Data Speed and Capacity	6G provides faster speeds, lower latency, and greater capacity, but extending these advantages beyond urban and affluent areas demands strategic planning and investment in expanding network infrastructure.
Energy Efficiency	Reducing energy consumption in devices and network infrastructure is crucial, particularly in regions with limited access to reliable power sources.
Digital Literacy and Education	Educating and empowering communities through digital literacy and training programs is essential for effective use of 6G technologies.
Privacy and Security	Implementing strong privacy and security measures is crucial, including safeguarding personal data and securing networks to mitigate cyber threats, particularly in underserved communities.
Regulatory and Policy Frameworks	Frameworks aim to tackle challenges such as spectrum allocation [64, 65], licensing, and universal service obligations to guarantee fair access to 6G technologies.
Customized Solutions for Specific Needs	Different regions or communities often face distinct challenges that need customized approaches.
Collaboration and Partnerships	Bridging the digital divide demands collective efforts from governments, private sectors, NGOs, and international organizations. Collaborative partnerships harness their combined expertise and resources to address underserved communities efficiently.

Table 3.
Digital divide.

new opportunities in transport and logistics. Transportation has an environmental perspective in which fuel consumption and greenhouse emissions gases are reduced. This will help create more sustainable and environmentally friendly transportation systems. The smart cities will also potentially be impacted [63] with regard to high-speed connectivity, IoT integration, improved AI and ML, immersive experiences, sustainable development and global connectivity, as depicted in **Table 3**. High-speed connectivity is essential for applications like smart traffic management, autonomous vehicles and remote healthcare. Furthermore, the IoT integration will bring efficiency in services like energy distribution, environmental monitoring and waste management. IoT integration will also seek improved AI and ML allowing efficient resource management, traffic optimization and improved public safety. Public safety can also be improved through more immersive experiences and interactive city planning. Ultimately, sustainability is enabled by intelligent energy networks, optimized energy management, and minimizing the carbon footprint of urban regions [66].

5. Challenges and considerations

B5G and 6G could be the next generation of wireless communication technologies. As for B5G some fundamental issues such as advanced infrastructure, spectrum availability, power consumption, security and privacy still need to be attended to offer improved quality of service (QoS) compared to 5G [67, 68]. There are special considerations that are crucial to the realization of B5G/6G applications and innovations.

These considerations will be divided into two categories. The first category, referred to as 'general,' will be addressed in Section 5.1. The second category, which includes economic, environmental, and ethical issues, will be detailed in Section 5.2, while the challenges will be presented in Section 5.3.

5.1 General considerations

The adoption of 6G technology is expected to satisfy the changing needs of society by opening up new applications and improving the quality of wireless communication. Ongoing research and innovation are essential to unlocking the full potential of 6G systems, bringing significant benefits to individuals, businesses, and industries alike. The following are future trends and considerations for 6G systems: efficient utilization of infrastructure resources, automated operations enabling service innovations, architectural evolution, service innovations and architecture application [69].

- *Efficient utilization of infrastructure resources*: Infrastructure resource efficiency is realized through cloud-native deployments that decouple from legacy silo architectures, exemplified by core network functions, radio access network (RAN), mobile edge computing and network [60] use cases supported by an integrated foundational platform with enhanced acceleration features.

- *Automated operations enabling service innovations*: automated operations that enable service innovations cover unified automation and operations and explain infrastructure programmability *via* network APIs.

- *Architecture evolution*: The inclusion of 6G control plane innovations within an evolved 5G core, with a simultaneous extension of service-based architecture to the RAN control plane [69].

- *Service innovations:* A service-oriented RAN/core control plane, decentralized non-access stratum, in-network processing, and the incorporation of computing capabilities at the telco edge, expanding on in-network and on-device computing.

- *Architecture application*: Reduced RAN/core deployment options, elimination of overlapping RAN/core functions (such as paging and handover), removal of certain 3GPP-specific protocols (like NG-AP and SCTP), utilization of programmable infrastructure, and integration of programmable all-photonics networks with data center infrastructure.

5.2 Economic, environmental and ethical considerations

With each generation of technological advancement, sustainability must be prioritized. This means carefully considering the economic, environmental, and ethical impacts of new technologies to ensure a sustainable future [70]. For 6G to truly benefit society, we must embed societal, environmental, and economic considerations into its entire lifecycle, from initial concept and design through deployment and beyond [71, 72]. Clearly, collaboration among researchers, policymakers, industry experts, and community representatives is crucial for responsible 6G development. This multidisciplinary approach will ensure that technological advancements align with broader societal goals and values. Furthermore, new design approaches are needed to account for the varying scopes [73] and scales of innovation and their contextual influence on perceived value.

5.3 Economic issues

1. *Investment and Infrastructure*: Clearly, the development and deployment of 6G technology will require significant investment in research, development, and infrastructure upgrades. 6G technology could spark innovation, attract investment, grow adoption, and revitalize telecommunications [74]. The transition to 6G requires significant investments with regard to the infrastructure. New towers will be required to provide broader coverage and higher bandwidth than 6G requires. Furthermore, optic fiber cables will be needed to carry the massive amounts of data generated by 6G. New software will also need to manage a complex network of 6G devices [75]. In August 2022, the National Science Foundation (NSF) announced a $100 million investment in the research and development of 6G. This funding will support various initiatives, including the creation of testbeds, collaborations with industry partners, and the exploration of novel architectural designs [76]. For progress's sake, five key areas such as driving value for both customers and the business, building versatile platforms, expanding the roster of network investors, making smart investments in digital infrastructure, and getting the right talent mix, need to be actioned by operators. In driving value for customers and business research and development efforts must focus on addressing operational issues Mobile network Operators (MNO) face as well as the issues hindering enterprises from scaling the use of advanced connectivity in their operation. Furthermore, by opening up new revenue sources, allowing MNOs to sell data produced by sensing could enhance the 6G investment's economics. Moreover, Strategic investments in digital infrastructure, such as accelerating xRAN rollout, upgrading fiber backhaul and fronthaul, [77] and investing in green energy, will support future 6G capital expenditures. Ultimately, developing the necessary talent pool with expertise in new network technologies is crucial for 6G success. Early investment in 6G talent acquisition will enable operators to capitalize on emerging opportunities.

2. *Economic Growth*: 6G, the next generation of mobile technology, presents a unique opportunity to reshape the world as we know it. With technology now at the heart of global competition and economic growth, 6G will define the international economic landscape for 2030 and beyond. Its impact will extend far beyond a simple technological advancement, driving breakthroughs across nearly every industry and facet of society [78].

3. *Digital Divide*: It is important to ensure that the benefits of 6G are accessible to everyone, regardless of their location or socioeconomic status. This will require addressing the digital divide and ensuring that affordable access is available to all. A significant digital divide persists globally, with many rural and remote areas lacking adequate connectivity. Low population density, low average incomes, challenging terrain, and absent infrastructure (like power grids) often make these areas unattractive for investment in connectivity networks [79]. Clearly, the digital divide remains a significant challenge, despite long-standing efforts from governments, NGOs, telecom and internet providers and other technology

players. In 2023, 2.6 billion people, a third of the world's population, lacked internet access, according to data from the International Telecommunications Union (ITU). Some of these people are in rural and remote parts of developed countries, but most are in the global south, where connectivity is a major hurdle [80]. Digitalization, powered by high-quality connectivity, is central to more equal access to healthcare, education, and employment, and can enable small businesses to participate in the digital economy. The digital divide hurts those communities who need the opportunities and economic prosperity offered by technology the most. Hence, bridging the digital divide is fundamental to sustainable development and the considerations in **Table 3** may assist.

5.4 Environmental

The 6G industry bears primary responsibility for minimizing the environmental impact of its goods, networks, and services throughout their lifecycle [81, 82]. To encourage sustainable practices and reduce environmental impact [81], organizations should be guided by regulatory frameworks (e.g., carbon pricing, emission standards, and environmental legislation) and focus on five key impact areas (**Table 4**): greenhouse gas and other emissions, energy, recycling and waste, water usage [83, 84], and land and biodiversity [72, 81].

Greenhouse Gas emissions: The manufacturing and disposal of electronic equipment for 6G networks is probably going to have a major impact on biodiversity loss, e-waste, and the depletion of natural resources. Furthermore, it is anticipated that the infrastructure needed to support 6G networks and the manufacturing of associated devices will raise greenhouse gas (GHG) emissions, worsening climate change. More research is required to fully comprehend the environmental impact of services and devices across their entire life cycle, despite their growing environmental consciousness. Building on these environmental concerns, the operational problems connected with 6G are mostly related to natural resource and energy use [82].

Energy Consumption: Energy efficiency in networks requires minimizing consumption during idle periods. AI-driven energy savings depend on this load-to-cost relationship, achievable only if low traffic translates to low energy states. Network observability is crucial, but its energy cost must be factored into AI optimizations. Furthermore, adapting network performance to fluctuating renewable energy availability will be key, especially as electrification increases demand. While low-carbon electricity reduces environmental impact, energy efficiency remains vital for cost-effectiveness.

Performance Indicator	Remarks
Green House Gas emissions	Emissions likely to influence climate change
Energy Consumption	Energy efficiency
Recycling and waste	E-waste will pose risk to humans and environment
Water usage	Efficient water management vital
land and biodiversity	Carbon sequestration, land use, and sustainability

Table 4.
Key performance indicators (KPI).

Recycling and Waste: One of the most important environmental issues of the twenty-first century is e-waste, or electronic garbage. E-waste management trends have changed throughout time, reflecting increased awareness, inventiveness, and the pressing need for long-term fixes [85]. The management of electronic rubbish, or "e-waste," was significantly affected by the rapid turnover of electronic equipment and technical obsolescence. We forecast that the number of abandoned smartphones, tablets, and other electronic gadgets will increase dramatically as people switch to 6G-enabled handsets, leading to an increase in e-waste generation. Numerous hazardous substances, including lead, mercury, cadmium, and brominated flame retardants, that are present in electronic equipment can pose a major risk to human health and the environment if improperly handled [86]. Additionally, improper e-waste burning, or landfilling contributes to soil and air pollution, which worsens environmental degradation and climate change [87]. The use of 6G technology in waste management allows for more efficient recycling and waste reduction strategies contributing to a healthier environment.

Water Usage: 6G-enabled manufacturing enhances energy and water management, reduces emissions, and promotes renewable energy use. Ericsson's Smart 5G Factory demonstrates this, with 5% waste reduction, 5% energy cost savings, and 24% increased energy efficiency [88] *land and biodiversity*: As indicated in **Table 5**, land and biodiversity are governed by factors such as land use, carbon sequestration, habitat protection, and land sustainability management. 6G will facilitate smart city applications that optimize land utilization and minimize urban sprawl, aiding in the preservation of natural habitats and reducing the adverse effects of urbanization on biodiversity. Furthermore, 6G will facilitate the monitoring and management of carbon sequestration initiatives, including reforestation and afforestation projects. This will aid in mitigating climate change and protecting ecosystems. Moreover, improving the capacity to monitor and protect biodiversity through advanced data analytics and AI-driven insights will effectively identify and address threats to biodiversity [89].

Ethical consideration	Remarks
Privacy and Data Protection	Protecting individual privacy requires safeguarding personal data, ensuring informed consent, and implementing strong data protection measures.
Fairness and Bias in AI Algorithms	Addressing biases in data, algorithms, and decision-making is crucial to prevent discrimination and ensure equitable outcomes.
Explainability and Transparency	Explainable AI empowers users by providing transparency into data usage and decision-making processes.
Using Emerging Technologies Ethically	Emerging technologies like augmented reality, virtual reality, and sophisticated machine learning algorithms should be implemented and used sensibly, without harming people or going against moral standards.
Economic and Social Implications	The social and economic effects on people and communities should be taken into account when developing 6G networks. To guarantee a fair and just transition, it is essential to evaluate the possible effects on employment, inequality, and socioeconomic gaps.
Inclusive and Accessible Connectivity	. Bridging the digital divide, ensuring affordability, and removing barriers to access are crucial for promoting equal opportunity and social inclusion.
Cybersecurity and Security	It is crucial to defend user data, network infrastructure, and vital systems against online attacks to ensure trust, privacy, and network integrity.

Table 5.
Ethical considerations.

5.5 Ethical considerations

As 6G network development advances, it is essential to address the ethical considerations and potential impacts of this transformative technology [90]. In technology development, ethical considerations address the possible dangers, damages, biases, and societal repercussions of technology and make sure that its design, implementation, and use are in line with moral principles and values, as explained in **Table 5**.

5.6 Challenges

- *Advanced infrastructure*: The advent of 6G is a function of advanced infrastructure, including the deployment of terahertz frequencies, massive fiber backbones, and a dense network of base stations and small cells. Notably, for rural and underserved areas, building infrastructure will attract high CAPEX and will be time-consuming.

- *Spectrum availability*: More spectrum is needed for 6G to deliver pertinent services. There are regulatory issues with the terahertz frequencies needed for 6G. In order to prevent interference and guarantee fair access, governments and regulators will need to assign new spectrum bands [4, 91] for 6G, which may require drawn-out legal procedures and international cooperation.

- *Power consumption*: This is a big worry because 6G will connect many devices at the same time. The sustainability of 6G networks will depend on how energy-efficient they are. Methods to lower the power consumption of the network infrastructure and the devices connected to it are still being researched.

- *Security*: The innovations for B5G wireless systems exhibit some similar traits. Firstly, the provision of level time, frequency and space adaptability is not consistent with past commercial systems. The second trait is that of a degree of flexibility exceeding previous implementations. These traits act in concert to provide an attack surface that is orders of magnitude more difficult to secure than previous wireless networks for 5G and beyond technologies [59, 92–94]. Generally, all layers of the network operate in a meticulously coordinated manner. In the past, much of this interaction was hardwired; however, both now and in the future, attackers can exploit vulnerabilities in any layer of the radio network, potentially triggering cascading effects across other layers.

6. Conclusion

Ever since the inception of wireless technology, wireless communication evolution has been a remarkable testament to human ingenuity. From the early days of radio transmission to the sophisticated 5G networks of today, each generation has brought about transformative advancements, reshaping the fabric of human society. 5G has ushered in an era of unprecedented connectivity, empowering industries, revolutionizing healthcare, and transforming how we interact and experience the world. B5G promises a further evolution which will usher in an era of hyper-connectivity. Clearly, given the potential to deliver unprecedented speeds, ultra-low latency and ubiquitous coverage, B5G will unpack a plethora of new applications ranging from autonomous systems and the Internet of Everything to immersive virtual and augmented reality experiences. However, developing and deploying B5G comes with significant challenges. It will be crucial to address concerns related to energy efficiency, security, privacy, and the digital divide to ensure that B5G's benefits are equitably shared and its development aligns with societal values and environmental sustainability. It is feasible to harness the power of future generations of wireless networks to create a more connected, equitable, and sustainable future for everyone by embracing innovation, addressing the challenges, and promoting responsible development. Wireless communication has enormous potential to change society in profound ways in the future.

Author details

Sindiso M. Nleya[1*], Mthulisi Velempini[1] and Tatenda T. Gotora[2]

1 University of Limpopo, Polokwane, South Africa

2 Postal and Telecommunications Authority of Zimbabwe (Potraz), Harare, Zimbabwe

*Address all correspondence to: sindiso.nleya@ul.ac.za

References

[1] Ivanova E, Iliev T, Stoyanov I, Mihaylov G. Evolution of mobile networks and seamless transition to 5G. IOP Conference Series: Materials Science and Engineering. 2021;**1032**(1):012008

[2] Lan NTM. Evolution of wireless technology: From 1G to 5G. Asian Journal of Applied Science and Technology. 2023;7(4):68-73

[3] Abu-Taieh E, Al Hadid IH, Zolait A. 5G road map to communication revolution. In: Cyberspace. London, UK: IntechOpen; 2020

[4] Phaswana P, Nleya SM, Velempini M. Centralized multicasting AODV routing protocol optimized for intermittent cognitive radio ad hoc networks: Multicasting AODV routing in intermittent cognitive radio ad hoc networks. In: EAI Endorsed Transactions on Mobile Communications and Applications. Vol. 8. European Alliance for Innovation (EAI); 2023

[5] Lin X, Lee N. 5G and Beyond. Cham, Switzerland: Springer Nature Switzerland AG; 2021

[6] Mateusz B. The Impact of Wi-Fi Technology on Modern Communication and Society. aithor.com; 2024 [Online]. Available from: https://aithor.com/essay-examples/the-impact-of-wireless-network-technology-on-modern-communication [Accessed: January 04, 2025]

[7] Wang C-X, Di Renzo M, Stanczak S, Wang S, Larsson EG. Artificial intelligence enabled wireless networking for 5G and beyond: Recent advances and future challenges. IEEE Wireless Communications. 2020;**27**(1):16-23

[8] Paul B. The "Beyond 5G/6G Era": A New Era of 7G Wireless Communication. paulbudde.com website; 2023 [Online]. Available from: https://paulbudde.com/blog/digital-media-apps-services/the-beyond-5g-6g-era-a-new-era-of-7g-wireless-communication/ [Accessed: January 03, 2025]

[9] Monique C. 6G: Shaping a Sustainable, Equitable Future. Martel-innovate website; 2024 [Online]. Available from: https://martel-innovate.com/news/2024/10/14/6g-shaping-a-sustainable-equitable-future/ [Accessed: January 03, 2025]

[10] Petri A. Societal Impact of 5G and 6G. Traficcom website; 2022 [Online]. Available from: https://www.traficom.fi/sites/default/files/media/file/013_5G%20Suomessa%20nyt%20ja%20tulevaisuudessa_Ahokangas.pdf [Accessed: January 04, 2025]

[11] Akbar MS, Hussain Z, Ikram M, Sheng QZ, Mukhopadhyay S. On challenges of sixth-generation (6G) wireless networks: A comprehensive survey of requirements, applications, and security issues. Journal of Network and Computer Applications. 2024;**233**:104040

[12] You X, Wang C-X, Huang J, Gao X, Zhang Z, Wang M, et al. Towards 6G wireless communication networks: Vision, enabling technologies, and new paradigm shifts. Science China Information Sciences. 2021;**64**:1-74

[13] Saurabh S, Kaushik S. Examining the impact of 6G telecommunications on society what to consider before the next generation of connectivity. IEEE Spectrum. 2023 [Online]. Available from: https://spectrum.

ieee.org/examining-the-impact-of-6g-telecommunications-on-society [Accessed: December 17, 2024]

[14] Agrawal R. Comparison of different mobile wireless technology (from 0g to 6g). ECS Transactions. 2022;**107**(1):4799

[15] Kumar P et al. Review paper on development of mobile wireless technology. Journal of Physics: Conference Series. 2021;**1979**(1):012024

[16] Hodara H, Skaljo E. From 1G to 5G. Fiber and Integrated Optics. 2021;**2-3**:85-183

[17] Frauendorf JL, de Souza A. The different architectures used in 1G, 2G, 3G, 4G, and 5G networks. In: The Architectural and Technological Revolution of 5G. Springer Nature Switzerland; 2022. pp. 83-107

[18] Javed AR, Abid R, Aslam B, Khalid HA, Khan MZ, Alhazmi OH, et al. Green5G: Enhancing capacity and coverage in device-to-device communication. Computers, Materials & Continua. 2021;**67**(2):1934-1950

[19] Abdullah A. The Evolution of Connectivity: Understanding the Power of 5G Technology. Linkedln; 2023 [Online]. Available from: https://www.linkedin.com/pulse/evolution-connectivity-understanding-power-5g-abdullah/ [Accessed: January 02, 2025]

[20] Yew HT, Dasuki A, Asmat AE, Japarudin MS, Herifian MNH, Mamat M. Internet-of-Things: Role of 5G wireless communications. In: 2023 12th International Conference on Awareness Science and Technology (iCAST). Chaoyang University of Technology, Taichung, Taiwan: IEEE; 2023

[21] Khanh QV, Hoai NV, Manh LD, Le AN, Jeon G. Wireless communication technologies for IoT in 5G: Vision, applications, and challenges. Wireless Communications and Mobile Computing. 2022;**2022**(1):3229294

[22] Rojek I, Kotlarz P, Dorożyński J, Mikołajewski D. Sixth-generation (6G) networks for improved machine-to-machine (M2M) communication in Industry 4.0. Electronics. 2024;**13**(10):1832

[23] Shah AS. A survey from 1G to 5G including the advent of 6G: Architectures, multiple access techniques, and emerging technologies. In: 2022 IEEE 12th Annual Computing and Communication Workshop and Conference (CCWC). Las Vegas, USA: IEEE; 2022

[24] Parashar J, Jain A, Meena L, Kapoor S. 5G network security challenges and solution. Journal of Intelligent Systems and Computing. 2023;**4**(2):1-13

[25] Ji S, Mishra AK. 5G security issues challenges and solutions against DDoS attacks: A survey. In: 2024 2nd International Conference on Disruptive Technologies (ICDT). IEEE; 2024

[26] Gupta M, Jha RK. Security Challenges in 6G Networks: A Game Theory-Based Intrusion Detection with Network Slicing and AI/ML Perspectives. Preprints. 2024

[27] Mishra R. An overview of backbone technology behind the latest advanced gadgets in use: 4G & 5G. Spatial Information Research. 2023;**31**(1):15-26

[28] Nadeem L, Azam MA, Amin Y, Al-Ghamdi MA, Chai KK, Khan MFN, et al. Integration of D2D, network slicing, and MEC in 5G cellular networks: Survey and challenges. IEEE Access. 2021;**9**:37590-37612

[29] Iqbal MU, Ansari EA, Akhtar S, Khan AN. Improving the QoS in 5G HetNets through cooperative Q-learning. IEEE Access. 2022;**10**:19654-19676

[30] Sharma S, Deivakani M, Reddy KS, Gnanasekar A, Aparna G. Key enabling technologies of 5G wireless mobile communication. Journal of Physics: Conference Series. 2021;**1817**:012003

[31] Jiang W, Han B. Evolution to first-generation (1G) Mobile cellular communications. In: Cellular Communication Networks and Standards: The Evolution from 1G to 6G. Springer Nature Switzerland AG; 2024. pp. 7-21

[32] Muppavaram K, Govathoti S, Kamidi D, Bhaskar T. Exploring the generations: A comparative study of mobile technology from 1G to 5G. International Journal of Electronics and Communication Engineering. 2023;**10**(7):54-62

[33] Richard G. CENGN. 2020 [Online]. Available from: https://www.cengn.ca/information-centre/innovation/timeline-from-1g-to-5g-a-brief-history-on-cell-phones/ [Accessed: December 21, 2024]

[34] Rajiv B. Evolution of Wireless Technologies 1G to 5G in Mobile Communication. RFpage; 2023 [Online]. Available from: https://www.rfpage.com/evolution-of-wireless-technologies-1g-to-5g-in-mobile-communication/ [Accessed: December 21, 2024]

[35] Pankaj S. Evolution of Mobile wireless communication networks-1G to 5G as well as future prospective of next generation communication network. International Journal of Computer Science and Mobile Computing. 2013;**2**(8):47-53

[36] What are the differences between 2G, 3G, 4G LTE, and 5G networks? [Online]. Available from: https://rantcell.com/comparison-of-2g-3g-4g-5g.html [Accessed: December 22, 2024]

[37] What is 3G?. Wireless Logic [Online]. Available from: https://wirelesslogic.com/iot-glossary/what-is-3g/ [Accessed: December 22, 2024]

[38] Bhavesh H. An overview of 4G technology. Iconic Research and Engineering. 2018;**1**(9):383-387

[39] Telecomtrainer. Introduction to beyond 5G. Telecomtrainer; 2025 [Online]. Available from: https://www.telecomtrainer.com/introduction-to-beyond-5g/ [Accessed: March 16, 2025]

[40] Ken N. Beyond 5G: The New Era of Connectivity. NEC; 2023 [Online]. Available from: https://www.nec.com/en/global/solutions/5g/blog/beyond-5g-the-new-era-of-connectivity.html [Accessed: January 04, 2025]

[41] Lee B. Recent Trend of Hyper-Connected Communication Technologies in 5G and Beyond. North Korea: Energy Excellence & Smart City Lab, Incheon National University;

[42] Kiwa JF, Chimhundu H, Tsvere M, Nleya S. Developing an animated interactive digital game using computer tools in relevance to intangible cultural heritage with Shona as a vehicle. IOSR Journal of Computer Engineering. 2021;**21**:23

[43] Kiwa JF, Chimhundu H, Tsvere M, Nleya S. Saving the vanishing intangible and tangible cultural heritage through the development of an online panorama. IOSR Journal of Computer Engineering. 2021;**21**:53-59

[44] Hyperconectivity How does it Affect us on a Daily Basis? [Online]. Available from: https://atriainnovation.com/en/

blog/hyperconnectivity-how-it-affects-us-in-the-day-to-day/ [Accessed: January 02, 2025]

[45] Aslam MM, Du L, Zhang X, Chen Y, Ahmed Z, Qureshi B. Sixth generation (6G) cognitive radio network (CRN) application, requirements, security issues, and key challenges. Wireless Communications and Mobile Computing. 2021;**2021**(1):1331428

[46] Mohsan SAH, Mazinani A, Malik W, Younas I, Othman NQH, Amjad H, et al. 6G: Envisioning the key technologies, applications and challenges. International Journal of Advanced Computer Science and Applications. 2020;**11**(9):14-23

[47] Espin. Next 6G: Key Features, Use Cases, and Challenges of Tomorrow's Wireless Revolution [Online]. Available from: https://www.e-spincorp.com/next-6g-key-features-use-cases/ [Accessed: December 23, 2024]

[48] Chen S, Kang S. The dual iconic features and key enabling technologies of 6G. Engineering. 2023;**28**:7-10

[49] Yash. From 5G to beyond: The next wave of connectivity. 2025 [Online]. Available from: https://vocal.media/chapters/from-5-g-to-beyond-the-next-wave-of-connectivity [Accessed: January 07, 2025]

[50] Siddiky MNA, Rahman ME, Uzzal MS, Kabir H. A comprehensive exploration of 6G wireless communication technologies. Computers. MDPI. 2025;**14**(1):15

[51] Marzetta TL. Noncooperative cellular wireless with unlimited numbers of base station antennas. IEEE Transactions on Wireless Communications. 2010;**9**(11):3590-3600

[52] Yrjölä S, Ahokangas P, Matinmikko-Blue M. Future scenarios and anticipated impacts of 6G. In: The Changing World of Mobile Communications: 5G, 6G and the Future of Digital Services. Cham, Switzerland: Springer International Publishing; 2023. pp. 5-92

[53] Zia U. The Future of 5G and beyond: A Hyper-Connected World. linkedin; 2024 [Online]. Available from: https://www.linkedin.com/pulse/future-5g-beyond-hyper-connected-world-zia-ullah-pa3df/ [Accessed: December 10, 2024]

[54] Ankush K. The Future Is Here: How 6G Technology Will Revolutionize Industries and Create New Opportunities. yash; 2023 [Online]. Available from: https://www.yash.com/blog/how-6g-technology-will-revolutionize-industries-and-create-new-opportunities/ [Accessed: January 07, 2025]

[55] Saurabh S. Unlocking the potential of 6G. In: How Ultra Fast Connectivity Is Set to Drive Business Innovation. appinventiv; 2024 [Online]. Available from: https://appinventiv.com/blog/6g-technology-transforming-enterprises/ [Accessed: January 08, 2025]

[56] Shen X, Gao J, Li M, Zhou C, Hu S, He M, et al. Toward immersive communications in 6G. Frontiers in Computer Science. 2023;**4**:1068478

[57] The applications of 6G: what will it change? [Online]. Available from: https://livingtomorrow.be/insights-and-stories/the-applications-of-6g-what-will-it-change [Accessed: January 6, 2025]

[58] 6G Technology? Future of Comm. simnovus; 2020 [Online]. Available from: https://simnovus.com/what-is-6g/#Personalized_Network_Experience [Accessed: January 07, 2025]

[59] Liu K, Li Q, Zhang H. Analysis of the impact of remanufacturing process innovation on closed-loop supply chain from the perspective of government subsidy. Sustainability. 2022;**14**(18):11333

[60] Divyanjali G. 6G and the Future Customer Experience. MantraLabs; 2022 [Online]. Available from: https://www.mantralabsglobal.com/blog/6g-and-the-future-customer experience/ [Accessed: January 07, 2025]

[61] Silvia R, Hamed A, Gordana G, Yoram H, Konstantinos K, Paolo G, et al. 5G and beyond networks. In: Inclusive Radio Communications for 5G and Beyond. Cambridge, MA, USA: Academic Press; 2021. pp. 141-186

[62] Alliance NG. 6G: The Next Frontier of Innovation and Investment. Washington, D.C., USA: Next G Alliance; 2023 [Online]. Available from: https://nextgalliance.org/white_papers/6g-next-frontier-innovation-investment/ [Accessed: January 07, 2025]

[63] Zina C, Tomas L, Fabian Q, Martin W, Information Technology Industry Council (ITI). Comments of the Information Technology Industry Council in Response to the Public Consultation on the Update of the Brazilian Strategy for Digital Transformation. Washington, D.C., USA: Information Technology Industry Council; 2024. Available from: www.itic.org

[64] Reza V-G. The Evolution of Fronthaul Networks. San Jose, California, USA: Viavi Solutions; 2020 [Online]. Available from: https://www.viavisolutions.com/en-us/literature/evolution-fronthaul-networks-white-papers-books-en.pdf [Accessed: January 04, 2025]

[65] Glenn R. 6G: The Next Frontier of Innovation and Investment. Arlington, Virginia, USA: Nextgalliance; 2023 [Online]. Available from: https://nextgalliance.org/white_papers/6g-next-frontier-innovation-investment/#:~:text=Technology%20has%20become%20the%20epicenter,landscape%20for%202030%20and%20beyond [Accessed: February 05, 2025]

[66] Mapfre. Impact of 5G and 6G Technologies on Healthcare. Madrid, Spain: Mapfre; 2023 [Online]. Available from: https://www.mapfre.com/en/insights/innovation/impact-5g-and-6g-technologies-on-healthcare/ [Accessed: February 06, 2025]

[67] Ahmed B. The Future of Connectivity: Understanding 6G and its Potential Impact on Technology and Applications. Linkedln; 2024 [Online]. Available from: https://www.linkedin.com/pulse/future-connectivity-understanding-6g-its-potential-impact-banafa-k4shc/ [Accessed: January 03, 2025]

[68] Medicalexpo. 6G in Healthcare: Next-Generation Connectivity. MedicalExpo e-magazine; 2024 [Online]. Available from: https://emag.medicalexpo.com/6g-in-healthcare-next-generation-connectivity [Accessed: February 05, 2025]

[69] Jha AV, Appasani B, Khan MS, Eadally S, Katib I. 6G for intelligent transportation systems: Standards, technologies, and challenges. Telecommunication Systems. 2024;**86**(2):1-28

[70] Apeksha Telecom. Ethical Considerations in the Development of 6G. India: Telecomgurukul; 2024 [Online]. Available from: https://www.telecomgurukul.com/post/ethical-considerations-in-the-development-of-6g#:~:text=By%20addressing%20

ethical%20considerations%20 such,and%20deployed%20 responsibly%20and%20ethically [Accessed: February 08, 2025]

[71] Sharma S, Popli R, Singh S, Chhabra G, Saini GS, Singh M, et al. The role of 6G Technologies in Advancing Smart City Applications: Opportunities and challenges. Sustainability. 2024;**16**(16):7039

[72] Supriya K. 6G Networks: Timeline, Features, and Potential Impact. Telecomgurukul; 2025 [Online]. Available from: https://www.telecomgurukul.com/post/6g-networks-timeline-features-and-potential-impact [Accessed: February 08, 2025]

[73] Ayoola VB, Idoko IP, Eromonsei SO, Afolabi O, Apampa AR, Oyebanji OS. The role of big data and AI in enhancing biodiversity conservation and resource management in the USA. World Journal of Advanced Research and Reviews. 2024;**23**(2):1851-1873

[74] Kadir EA, Shubair R, Rahim SKA, Abdul SK, Himdi M, Kamarudin MR, et al. B5G and 6G: Next generation wireless communications technologies, demand and challenge. In: 2021 International Congress of Advanced Technology and Engineering (ICOTEN). Piscataway, NJ, USA: IEEE; 2021

[75] Sean K. Five Core Network Design Considerations from 5G to 6G. USA: RCR Wireless; 2024 [Online]. Available from: https://www.rcrwireless.com/20240416/fundamentals/five-core-network-design-considerations-from-5g-to-6g [Accessed: February 08, 2025]

[76] Bhushan J, Gagandeep B, Ralf B, Ian D. Shaping a Greener Future: The Role of Environmental Sustainability in 6G Technology – Next G Alliance. Washington, D.C., USA: nextgalliance; 2024 [Online]. Available from: https://nextgalliance.org/shaping-a-greener-future-the-role-of-environmental-sustainability-in-6g-technology/ [Accessed: February 04, 2025]

[77] Chauhan D, Mewada H, Gondalia V, Almalki FA, Patel S, Modi H, et al. Balancing technological innovation and environmental sustainability: A lifecycle analysis of 6G wireless communication technology. Sustainability. 2024;**15**(16):6533

[78] Zina C, Tomás L, Fabian Q, Martin W. Shaping the Future of 6G. New York City, USA: McKinsey; 2024 [Online]. Available from: https://www.mckinsey.com/industries/technology-media-and-telecommunications/our-insights/shaping-the-future-of-6g [Accessed: February 03, 2025]

[79] Roger K. 6G on the Horizon: Redefining Wireless Possibilities. Lisle, Illinois, USA: Molex [Online]. Available from: https://www.molex.com/en-us/blog/6g-on-the-horizon; [Accessed: February 04, 2025]

[80] Hemant K. The Future of Connectivity: Exploring the US 6G Market. Linkedin; 2023 [Online]. Available from: https://www.linkedin.com/pulse/future-connectivity-exploring-us-6g-market-hemant-kolhe-ak4wf/ [Accessed: February 02, 2025]

[81] Harri S. 6G White Paper on Connectivity for Remote Areas. Finland: University of Oulu; 2020

[82] Subho M. Innovating to Bridge the Digital Divide. Espoo, Finland: Nokia; 2024 [Online]. Available from: https://www.nokia.com/blog/innovating-to-bridge-the-digital-divide/ [Accessed: February 03, 2025]

[83] Nleya SM, Bagula A, Zennaro M, Pietrosemoli E. Optimisation of a TV

white space broadband market model for rural entrepreneur. Journal of ICT Standardization. 2014;**2**:109-128

[84] Gustav W. Why every Decision on 6G Must Put Sustainability First. Stockholm, Sweden: Ericsson; 2023 [Online]. Available from: https://www.ericsson.com/en/blog/2023/8/6g-sustainability-first [Accessed: February 06, 2025]

[85] Anna A, Katrina P, Margot B. G and Sustainability: Shaping the Next Generation of Connectivity. 6G4Society; 2025 [Online]. Available from: https://6g4society.eu/2025/01/15/6g-and-sustainability-shaping-the-next-generation-of-connectivity/ [Accessed: February 06, 2025]

[86] Nleya SM, Velempini M. Industrial metaverse: A comprehensive review, environmental impact, and challenges. Applied Sciences. 2024;**14**(13):5736

[87] Ndlovu S, Andile D, Nleya SM, Dube SS, Zengeya T, Marabada N, et al. Web application for reading water meters. In: 2023 2nd Zimbabwe Conference of Information and Communication Technologies (ZCICT). Gweru: IEEE; 2022

[88] Shumba E, Ndlovu BM, Nleya S, Moyo N. An assessment of internet of things (IoT) adoption readiness in water and sanitation in Zimbabwe. A case of the City of Gweru. In: 2022 1st Zimbabwe Conference of Information and Communication Technologies (ZCICT). IEEE; 2022

[89] Angela O. Trends in e-waste management. Africa Sustainability Matters. 2024 [Online]. Available from: https://africasustainabilitymatters.com/trends-in-e-waste-management [Accessed: February 06, 2025]

[90] Ghulam ST, Abushammala H. Challenges and opportunities in the management of electronic waste and its impact on human health and environment. Sustainability. 2023;**15**(3):1837

[91] Nleya SM. Design and Optimisation of a Low Cost Cognitive Mesh Network. Cape Town, South Africa: University of Cape Town; 2016

[92] Satish D. Building a more Sustainable Future with 6G. Santa Rosa, California, USA: Keysight; 2023 [Online]. Available from: https://www.keysight.com/blogs/en/inds/cell/2022/09/13/building-a-more-sustainable-future-with-6g [Accessed: February 06, 2025]

[93] Ruzomberka E, Love DJ, Brinton CG, Gupta A, Wang C-C, Poor HV. Challenges and opportunities for beyond-5G wireless security. IEEE Security & Privacy. 2023;**21**(5):55-66

[94] Nleya SM, Velempini M, Gotora TT. Modelling lanchester style DDOS attack and defense. In: 2022 1st Zimbabwe Conference of Information and Communication Technologies (ZCICT), Harare. IEEE; 2022

Section 2

Technologies and Applications in 5G and Beyond

Chapter 3

5G and Beyond: Advancements in Wireless Communications for IoT and Smart Cities

Srinivas Ramavath, Umesh Chandra Samal, Prasanta Kumar Patra, Pattepu Sunil and Bhargav Appasani

Abstract

This chapter has explored the role of next-generation wireless communication technologies, particularly 5G and Beyond 5G (B5G), in the context of the Internet of Things (IoT) and smart cities. The chapter will address how advanced wireless communication frameworks enable the efficient integration of IoT devices, ensuring low latency, high throughput, and reliable communication, which are essential for real-time applications like autonomous vehicles, smart grids, and urban infrastructure. Network slicing and edge computing technologies will also be discussed as key enablers for tailored service delivery in these environments. Moreover, the chapter will outline future directions, including the path toward 6G, and their potential impact on mobile networks and urban ecosystems.

Keywords: 5G, IoT, smart cities, wireless communication, network slicing

1. Introduction

The advent of 5G wireless technology has marked a transformative era in the field of communications [1], delivering unprecedented speeds, ultra-low latency, and massive connectivity [2]. Beyond 5G (B5G) and emerging 6G technologies promise to extend these capabilities, offering innovative solutions for complex challenges in the Internet of Things (IoT) and smart city applications [3]. 5G has revolutionized IoT by enabling real-time data exchange among billions of interconnected devices, fostering automation, precision, and efficiency across various sectors [4]. Smart cities leverage this technology to enhance urban living by optimizing transportation systems, energy management, public safety, and environmental monitoring [5]. The rapid evolution of wireless communication technologies has been a cornerstone of modern connectivity, fundamentally reshaping industries, economies, and daily life. With the advent of 5G, wireless networks have entered a new era of unprecedented speed, ultra-low latency, and massive device connectivity [6]. These advancements are particularly transformative for the Internet of Things (IoT) and smart city applications, enabling real-time data exchange, automation, and enhanced decision-making processes. As urbanization accelerates and the number of connected devices grows exponentially, the demand for high-performance wireless communication infrastructures has never been greater. 5G not only addresses this demand by offering enhanced mobile broadband (eMBB), ultra-reliable low-latency communications (URLLC), and massive machine-type communications (mMTC), but it also paves the way for future wireless technologies beyond 5G, such as 6G, AI-driven network optimization, and terahertz (THz) communication [7]. In smart cities, 5G empowers a range of innovative applications, including intelligent traffic management, autonomous vehicles, smart grids, environmental monitoring, and connected healthcare systems. These applications require seamless integration of edge computing, artificial intelligence (AI), and distributed networking to ensure efficiency, security, and reliability. Similarly, in IoT ecosystems, 5G facilitates large-scale machine-to-machine (M2M) communication, enabling smart factories, automated logistics, and precision agriculture [8]. Despite its advantages, 5G adoption also presents challenges, such as infrastructure deployment costs, security concerns, and spectrum allocation issues. Addressing these challenges will be crucial as researchers and engineers push the boundaries of wireless communication beyond 5G, exploring new paradigms like reconfigurable intelligent surfaces (RIS), satellite-terrestrial integration, and quantum communication [9].

This chapter delves into the advancements in wireless communication that fuel IoT and smart city innovations. It explores how 5G serves as a foundational enabler and what future technologies beyond 5G hold for the next generation of hyper-connected societies. Advancements beyond 5G focus on improving spectrum efficiency, network densification, and the integration of artificial intelligence (AI) to create intelligent, adaptive, and autonomous communication networks. These developments are pivotal in supporting futuristic applications such as holographic communication, autonomous vehicles, and ubiquitous sensing, laying the foundation for a seamlessly connected digital ecosystem. By addressing challenges like scalability, energy efficiency, and security, 5G and beyond are reshaping wireless communications and unlocking new possibilities for IoT and smart city innovations [10].

i. Overview of 5G technology and its role in IoT and smart cities.

ii. Evolution of wireless communication: from 1G to 5G and beyond.

iii. Key challenges in traditional wireless networks for IoT and smart cities.

Figure 1 illustrates the evolution of mobile network technology from 1G to 5G. Here is a breakdown: 1G (1980s). Speed: 2.4 Kbps. Technology: Analog. Features: Voice-only communication. 2G (1990s). Speed: 64 Kbps. Technology: Digital. Features: SMS (text messaging) introduced, better voice quality. 3G (2000s). Speed: 3.1 Mbps. Technology: Broadband. Features: Basic internet access, multimedia (music, video calling, emails). 4G (2010s). Speed**: 100 Mbps. Technology: High-speed internet. Features: Video streaming, online gaming, improved mobile browsing. 5G (2020s). Speed: 10 Gbps. Technology: Ultra-fast connectivity. Features: IoT (Internet of Things), smart cities, automation, AI-driven applications, and improved industrial connectivity. Each generation has brought advancements in speed, connectivity, and functionality, shaping modern communication and digital experiences. The evolution of mobile communications has been a remarkable journey, transforming the way people connect and share information [11]. It began with the introduction of 1G in

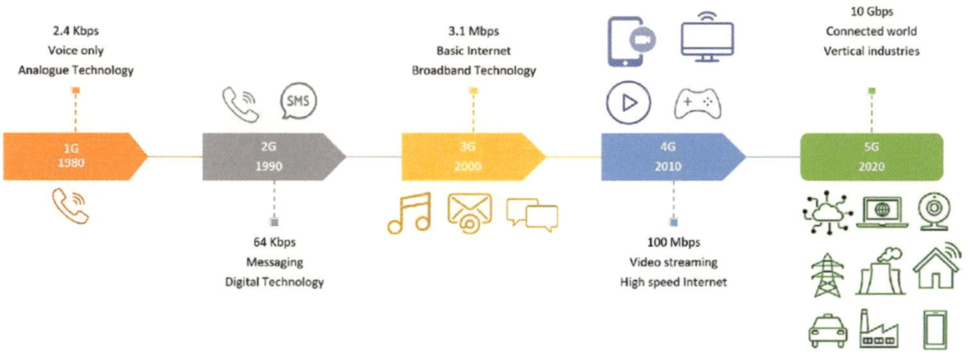

Figure 1.
Mobile communications evolution.

the 1980s, offering analog voice communication that was revolutionary for its time. This paved the way for 2G in the 1990s, which introduced digital encryption, text messaging (SMS), and better call quality [12]. The early 2000s saw the advent of 3G, bringing higher data speeds and enabling mobile internet access, which changed how people consumed media and interacted online. The launch of 4G in the 2010s marked a significant leap, offering high-speed data, seamless video streaming, and support for advanced applications like mobile gaming and video conferencing [13]. Now, with the rollout of 5G, mobile communications have entered an era of ultra-low latency, massive device connectivity, and blazing-fast speeds, enabling innovations in fields like IoT, autonomous vehicles, and smart cities [14]. Each generation has not only improved connectivity but has also reshaped industries and everyday life, underscoring the transformative power of mobile technology [15].

2. Foundations of 5G in IoT and smart cities

The foundation of 5G in IoT (Internet of Things) and smart cities lies in its transformative capabilities to connect devices, systems, and people seamlessly. Its architecture and features are designed to address the growing demands of hyper-connectivity, enabling applications critical to IoT ecosystems and urban development.

Figure 2 represents how 5G technology supports vertical industries in a smart city. It highlights different sectors and their respective applications of 5G connectivity, key industries, and applications of 5G in a smart city: 1. Energy, intelligent grid: Smart energy distribution, real-time monitoring, and automated grid management [2]. Distributed energy management: Efficient energy allocation from renewable sources like solar and wind. 2. Transport, intelligent transport systems: Connected and autonomous vehicles, smart traffic management, and real-time route optimization [16]. 3. Healthcare, telemedicine: Remote medical consultations, virtual healthcare, and AI-driven diagnostics. Remote monitoring: Wearable health devices for real-time tracking of patients' vitals and emergency alerts [17]. 4. Manufacturing, automation, and asset management: Smart factories, AI-driven automation, and real-time equipment tracking. Safety and quality assurance: AI and IoT-based quality control, workplace safety monitoring, and predictive maintenance [18]. 5. Media and entertainment, advertising: Personalized and AI-driven advertisements using real-time data and augmented reality. Immersive media: Virtual reality (VR), augmented reality

Figure 2.
Vertical industries in smart cities and their applications.

(AR), and enhanced digital experiences. Significance of 5G in smart cities, faster data speeds: Enables real-time communication and high-speed connectivity. Low latency: Ensures instant responses, which is crucial for healthcare and automation. Massive IoT connectivity: Supports smart devices and automation across industries. Energy efficiency: Optimizes power usage in grids and industries. 5G is revolutionizing smart cities by enhancing connectivity, efficiency, and automation across industries, leading to sustainable urban development. Smart cities leverage advanced technologies to improve urban living, enhance sustainability, and drive economic growth. Key vertical industries in smart cities include transportation, energy, healthcare, public safety, and water management. In transportation, smart cities use IoT-enabled traffic management systems, autonomous vehicles, and real-time public transit updates to reduce congestion and improve mobility. The energy sector benefits from smart grids, renewable energy integration, and demand-response systems that optimize energy consumption and reduce waste. Healthcare advancements include telemedicine, connected health monitoring devices, and AI-driven predictive analytics to enhance patient care and emergency response. Public safety relies on AI-powered surveillance, smart street lighting, and predictive policing to enhance security and reduce crime rates. In water management, technologies such as IoT-enabled sensors monitor water quality, detect leaks, and optimize distribution networks, ensuring efficient and sustainable usage. These verticals collectively foster innovation, improve quality of life, and contribute to the economic and environmental resilience of smart cities [8].

1. Characteristics of 5G (e.g., enhanced mobile broadband, ultra-reliable low-latency communication, massive machine-type communication).

2. Core technologies enabling 5G: Millimeter waves, small cells, massive Multiple Input, Multiple Output (MIMO), beamforming.

3. 5G network slicing for IoT and smart city applications.

DOI: http://dx.doi.org/10.5772/intechopen.1009925

2.1 The key elements underpinning 5G's role in these domains

1. *Key features of 5G for IoT and smart cities:* Enhanced data rates in 5G technology provide speeds of up to 10 Gbps, facilitating real-time data transmission for bandwidth-intensive applications such as high-resolution video surveillance, autonomous vehicular networks, and immersive virtual reality in smart city environments. Ultra-low latency in 5G networks, reaching as low as 1 ms, ensures reliable support for mission-critical IoT applications, including remote surgical procedures, autonomous transportation systems, and industrial automation, where instantaneous communication is essential. Massive connectivity capabilities allow 5G networks to support up to 1 million connected devices per square kilometer, making them well-suited for highly dense urban environments with extensive IoT deployments, such as sensor networks, smart infrastructure, and connected devices. Network slicing enables the dynamic creation of virtualized, application-specific network partitions, ensuring optimized performance for diverse use cases, including emergency response communications, intelligent energy grids, and next-generation entertainment services. Energy efficiency in 5G architecture is designed to reduce power consumption per transmitted bit, thereby enhancing the operational longevity of IoT devices, minimizing energy expenditures, and contributing to sustainable smart city development.

2. *IoT in the 5G ecosystem:* 5G serves as a fundamental enabler of the Internet of Things (IoT) by meeting the specific demands of diverse applications. Massive IoT (mIoT) is optimized for low-power, wide-area networks (LPWAN), facilitating large-scale deployments such as smart meters, environmental monitoring systems, and precision agriculture. Critical IoT leverages ultra-reliable, low-latency communication (URLLC) to support mission-critical applications, including industrial automation, medical devices, and autonomous transportation systems. Broadband IoT capitalizes on high-speed data transmission to enable advanced use cases such as real-time video analytics, augmented reality, and digital twin technologies for urban infrastructure planning.

3. *Role of 5G in smart cities:* In order to maximize resources, improve public services, and raise living standards, smart cities rely on linked networks. 5G lays the groundwork for these developments. Vehicle-to-everything (V2X) connectivity, intelligent traffic control systems, and autonomous mobility solutions are all made possible by 5G smart transportation. Management of energy: 5G guarantees effective energy distribution and consumption monitoring through smart grids and Internet of Things-enabled energy systems. Security and public safety at robust sensor networks, predictive policing, and real-time video monitoring are all made possible by high-speed connection. To assist sustainable urban growth, environmental monitoring in 5G-enabled IoT devices tracks water levels, trash management, and air quality. 5G's dependable and low-latency connectivity is advantageous for healthcare, including telemedicine, remote monitoring, and linked healthcare systems.

4. *Core technologies enabling 5G for IoT and smart cities:* Millimeter Waves (mmWave) provide enhanced bandwidth, making them well-suited for dense urban deployments and IoT applications that require high data rates. Massive MIMO (Multiple Input, Multiple Output) improves network capacity and

efficiency by enabling concurrent communication with multiple devices. Edge computing brings computational resources closer to end devices, thereby reducing latency and facilitating real-time analytics for IoT and smart city implementations. Artificial Intelligence (AI) and Machine Learning (ML) are integrated into 5G networks to enable dynamic resource allocation, predictive maintenance, and adaptive security mechanisms. Additionally, Software-Defined Networking (SDN) and Network Function Virtualization (NFV) enhance network management by increasing flexibility and scalability, thereby optimizing the deployment of IoT and smart city solutions.

5. *Challenges and considerations:* While 5G technology presents immense potential, its integration into IoT ecosystems and smart city infrastructure is accompanied by several challenges. High deployment costs: Establishing the necessary infrastructure, including base stations and fiber optic networks, requires substantial capital investment. Data security and privacy: The proliferation of connected devices increases the risk of data breaches, necessitating robust security measures to protect sensitive information. Interoperability: Achieving seamless communication among diverse IoT devices and legacy systems demands standardized protocols to ensure compatibility. Energy consumption: Although 5G devices are designed for efficiency, the overall network operation, particularly in densely populated regions, may result in significant energy demands. The foundation of 5G in IoT and smart cities is built on its ability to provide a reliable, high-performance, and scalable network. As urban centers continue to grow and the demand for IoT expands, 5G will play a pivotal role in shaping the future of connected living and smart city innovation. The role of 5G technologies in a smart city is the case for the introduction of intelligent transportation systems.

The rapid evolution of wireless communication technologies has paved the way for the development of smart cities, where the integration of advanced systems enhances urban living. Among these advancements, 5G technology emerges as a cornerstone for enabling a wide array of smart city applications. One of the most promising domains benefiting from 5G is the Intelligent Transportation System (ITS). This paper explores how 5G technologies contribute to the development and optimization of ITS in smart cities, addressing key challenges, use cases, and future implications.

2.2 The fundamentals of 5G in smart cities

5G technology, characterized by its ultra-low latency, high bandwidth, and massive device connectivity, offers significant improvements over its predecessors. These capabilities enable real-time data transmission and processing, which are critical for applications in a smart city ecosystem. In the context of ITS, 5G facilitates seamless communication among vehicles, infrastructure, and pedestrians, forming the foundation for a connected and efficient urban transportation network.

2.3 Key benefits of 5G for intelligent transportation systems

Enhanced Vehicle-to-Everything (V2X) communication: 5G supports Vehicle-to-Everything (V2X) communication, which includes Vehicle-to-Vehicle (V2V), Vehicle-to-Infrastructure (V2I), and Vehicle-to-Pedestrian (V2P) interactions.

These communication types ensure that vehicles can share real-time data about traffic conditions, potential hazards, and route optimization. Real-time traffic management: The high-speed data transfer capabilities of 5G enable dynamic traffic signal adjustments based on real-time traffic flow. This reduces congestion, improves road safety, and minimizes environmental impact. Autonomous vehicle enablement: 5G is critical for the operation of autonomous vehicles, as it ensures the ultra-reliable and low-latency communication required for real-time decision-making and navigation. Public transportation optimization: Smart busses, trains, and subways can utilize 5G to provide real-time updates to passengers, optimize routes based on demand, and improve overall service efficiency.

2.4 Use cases of 5G in ITS

Smart traffic signals: Adaptive traffic signals powered by 5G can analyze traffic patterns and adjust timings in real time, reducing wait times and improving fuel efficiency. Connected vehicle ecosystems: Vehicles equipped with 5G-enabled sensors can exchange information about speed, location, and road conditions, creating a safer and more efficient driving environment. Pedestrian safety systems: 5G-enabled systems can alert drivers and pedestrians to potential collisions, especially in densely populated urban areas. Smart parking solutions: 5G facilitates real-time updates on parking availability, guiding drivers to vacant spots and reducing time spent searching for parking.

2.5 Challenges and considerations

Despite its potential, the implementation of 5G in ITS faces several challenges in infrastructure costs. Deploying 5G networks requires significant investment in infrastructure, including base stations, antennas, and fiber optic connections. Data security and privacy: The vast amounts of data generated by 5G-enabled systems raise concerns about cybersecurity and data privacy. Interoperability is ensuring compatibility between different devices, platforms, and manufacturers is essential for a seamless ITS. Regulatory and policy barriers: Governments and regulatory bodies need to establish clear guidelines to support the adoption of 5G technologies in transportation systems.

2.6 Future implications

The integration of 5G into ITS marks a transformative shift in urban mobility. Future developments may include integration with AI and IoT: Combining 5G with artificial intelligence (AI) and the Internet of Things (IoT) will enable predictive analytics, proactive maintenance, and further automation of transportation systems. Sustainability goals: 5G-enabled ITS can contribute to reduced carbon emissions by optimizing traffic flow and encouraging the use of public transportation and electric vehicles. Smart city synergies: Beyond transportation, 5G will enable interconnected urban systems, including energy management, healthcare, and public safety. 5G technologies are poised to revolutionize intelligent transportation systems, making urban mobility safer, more efficient, and more sustainable. While challenges remain, the opportunities presented by 5G far outweigh the hurdles, promising a future where smart cities thrive through interconnected and intelligent systems. Policymakers, technologists, and urban planners must collaborate to harness the full potential of 5G, ensuring that ITS serves as a model for innovation and progress in the era of smart cities. 5G will have a transformative impact on smart city transport infrastructure, enabling more efficient, sustainable, and interconnected urban mobility systems. Here is how it could shape the future. Enhanced connectivity and data flow real-time data exchange: 5G's low latency and high bandwidth will facilitate instantaneous communication between vehicles, infrastructure, and control systems. This allows for real-time traffic monitoring, navigation updates, and safety alerts. IoT integration: A robust 5G network can support a vast number of Internet of Things (IoT) devices, such as smart traffic lights, connected sensors, and parking systems, to improve traffic flow and reduce congestion. Support for autonomous vehicles Vehicle-to-Everything (V2X) communication: 5G enables autonomous vehicles to communicate with each other (V2V), with infrastructure (V2I), and with pedestrians (V2P), enhancing safety and coordination. Reduced latency: The near-instantaneous response times of 5G are critical for the safe operation of self-driving cars, particularly in urban environments with complex traffic conditions. Improved public transport systems. Smart fleet management: 5G can improve the efficiency of public transport through real-time tracking and management of busses, trains, and other transit systems. Dynamic routing: Real-time data from 5G networks can enable public transport systems to adjust routes and schedules dynamically based on demand, weather, and traffic conditions. Optimized traffic management. Smart traffic lights: AI-driven traffic lights connected via 5G can adapt to real-time traffic patterns, reducing delays and emissions. Incident management: Faster communication allows for quicker response to accidents, road blockages, or emergencies. Enhanced user experience. Seamless connectivity for commuters: 5G enables passengers to enjoy uninterrupted high-speed internet on public transport, making commuting more productive and enjoyable. Personalized services: Integration of 5G with apps can provide users with tailored transit recommendations, including multimodal transport options. Sustainability and energy efficiency Electric Vehicle (EV) ecosystem: 5G can support smart charging infrastructure, optimizing energy usage and promoting electric mobility. Environmental monitoring: Sensors connected via 5G can monitor air quality and noise pollution, enabling data-driven urban planning. Data-driven decision-making advanced analytics: 5G facilitates the collection and analysis of vast amounts of data from transport systems, helping city planners design better infrastructure. Predictive maintenance: Real-time data can identify potential issues in infrastructure (e.g., bridges, tunnels, railways) before they become critical. By enabling faster, more reliable, and interconnected systems, 5G will be a cornerstone of the smart city transport revolution, making urban mobility safer, more efficient, and environmentally friendly.

3. Advancements beyond 5G (5G/6G)

Emerging technologies in B5G/6G: 1. Terahertz communication. 2. Artificial Intelligence (AI)-enabled wireless systems. 3. Intelligent reflecting surfaces (IRS). Quantum communication and its potential in smart cities. Integrated sensing and communication (ISAC). Advancements Beyond 5G (B5G/6G) represent the next phase of innovation in wireless communication technology, building on the foundation of 5G to offer significantly enhanced capabilities.

3.1 Overview of key aspects

Enhanced speed and capacity are expected to exceed 1 Tbps, enabling real-time transmission of massive datasets and ultra-high-definition media. Spectrum utilization: Use of sub-terahertz (THz) and terahertz frequencies to provide higher bandwidth and minimize congestion. Ultra-low latency target latencies as low as 0.1 ms, are crucial for applications like autonomous vehicles, robotic surgery, and industrial automation. Massive connectivity support for 1 million devices per square kilometer, accommodating the exponential growth of IoT devices in smart cities, agriculture, and healthcare. Intelligent network management incorporates AI and machine learning for dynamic resource allocation, predictive maintenance, and self-healing networks. Enhanced energy efficiency focuses on green technology to reduce energy consumption, utilizing AI for energy-efficient operations and advanced materials for energy harvesting. New applications and use cases immersive experiences: Holographic communication, augmented reality (AR), and virtual reality (VR) on a massive scale. Tactile internet: Real-time haptic feedback enabling remote control of devices with precision. Digital twins: Real-time digital replicas of physical entities for predictive modeling and optimization in industries. Integration of advanced technologies. Quantum communication: Incorporating quantum encryption for unparalleled security. AI at the edge: Enabling real-time decision-making on devices without reliance on centralized processing. Satellite integration: Seamless interconnectivity between terrestrial and non-terrestrial networks for global coverage. Global collaboration and standards: Development of international standards and regulations to ensure interoperability and fair access. Cross-sector partnerships to drive innovation in diverse industries like healthcare, transportation, and entertainment. Challenges to overcome spectrum availability: Managing interference and optimizing underutilized frequencies. Infrastructure development: Building the necessary hardware to support THz communications. Security is addressing vulnerabilities in hyper-connected environments. Ensuring affordability and accessibility to prevent a digital divide. While 6G networks are expected to roll out around 2030, research and early development are already underway. These advancements aim to transform communication into an intelligent, ubiquitous, and sustainable ecosystem.

4. IoT in the era of 5G and beyond

5G and B5G use cases in IoT: Smart homes and appliances. Industrial IoT (IIoT) and automation. Healthcare IoT applications. Connectivity requirements and challenges for IoT devices. The Internet of Things (IoT) is undergoing a transformative evolution with the advent of 5G and upcoming Beyond 5G (B5G) technologies. These advancements enable IoT devices to operate with unprecedented speed, reliability, and scalability, unlocking new possibilities across industries.

4.1 Key features and enhancements

Massive connectivity device density: 5G can support up to 1 million devices per square kilometer, a significant improvement over previous generations. Smart environments: Enables smart cities, connected homes, and industrial IoT (IIoT) with millions of devices interacting seamlessly. Ultra-low latency: 5G reduces latency to as low as 1 ms, while B5G/6G aims for sub-millisecond latency, critical for real-time applications like: Autonomous.

4.2 Vehicles, remote surgery, precision manufacturing

High-speed data transmission: With data rates up to 10 Gbps and beyond in B5G, IoT devices can handle massive data loads, enabling high-definition video streams for security cameras, real-time analytics in IoT ecosystems, enhanced augmented reality (AR) and virtual reality (VR) experiences, energy efficiency, and advanced protocols like sleep modes and energy harvesting in IoT devices to reduce power consumption. 5G networks prioritize energy-efficient communication, extend device battery life, and support green IoT initiatives.

4.3 Impact on key IoT applications

Smart cities' real-time traffic management using connected sensors and vehicles. Smart grids with dynamic energy distribution and fault detection. Enhanced public safety through connected surveillance and emergency systems. Healthcare remote patient monitoring using wearable IoT devices. Real-time diagnostics with AI-enabled medical sensors. Remote robotic surgeries are supported by ultra-low latency. Industry 4.0 predictive maintenance through sensor networks in factories. Autonomous robots and drones for logistics and supply chains. Real-time monitoring and optimization of industrial processes. Agriculture precision agriculture using connected sensors for soil, weather, and crop monitoring. Autonomous farming equipment for planting, harvesting, and irrigation. Livestock tracking with IoT wearables. Transportation vehicle-to-everything (V2X) communication for autonomous driving. Smart fleet management with real-time tracking and route optimization. Connected infrastructure for safer and more efficient travel.

4.4 Technological enablers

Edge computing reduces latency by processing data closer to IoT devices. Enhances real-time decision-making in critical applications. Network slicing allocates dedicated network resources for specific IoT applications. Ensures reliability and performance for mission-critical use cases. Artificial Intelligence (AI) integration enables predictive analytics and automated decision-making. Enhances anomaly detection and system optimization in IoT networks.

4.5 Challenges and considerations

Security and privacy: IoT devices are vulnerable to cyberattacks, necessitating robust security measures like encryption and authentication. Data privacy regulations must be adhered to as IoT networks expand. Scalability is managing billions of devices that require advanced protocols and efficient resource allocation. Cost and accessibility ensure affordability for small-scale users and developing regions, which are critical for widespread adoption. Standardization interoperability among diverse IoT devices and networks needs standardized protocols. The combination of IoT with 5G and Beyond (B5G/6G) will reshape industries, enhance human experiences, and pave the way for smart ecosystems. Key advancements such as ultra-reliable communication, intelligent automation, and ubiquitous connectivity will define the next generation of IoT applications, creating a seamlessly interconnected world.

5. Smart cities powered by 5G and beyond

Role of 5G in enabling smart city services: Intelligent transportation systems, smart grids and energy management, public safety and surveillance, integration of IoT with smart city infrastructure, case studies of 5G-enabled smart cities. The advent of 5G and Beyond (B5G/6G) technologies is revolutionizing the concept of smart cities, enabling interconnected ecosystems that enhance urban living, improve sustainability, and boost economic growth. These advanced networks provide the backbone for seamless connectivity, real-time data exchange, and intelligent decision-making.

5.1 Core features enabled by 5G and beyond

Ultra-high-speed and capacity support for up to 1 million devices per square kilometer allows massive IoT deployment, facilitating smart lighting, waste management, and environmental monitoring. High-speed data transfer enables real-time analytics for urban planning and emergency response. Ultra-low latency as low as 1 ms (and even lower in B5G/6G) supports critical applications like autonomous transportation systems and remote healthcare services. Reliability and network slicing: ultra-reliable low-latency communication (URLLC) ensures consistent performance for mission-critical services. Network slicing allows dedicated virtual networks for specific applications, such as public safety, healthcare, and industrial automation. Integration of AI and edge computing: AI-driven analytics optimize resource usage, predict maintenance needs, and improve citizen services. Edge computing reduces data transmission delays by processing information closer to the source.

5.2 Key applications of smart cities powered by 5G/B5G

Smart transportation autonomous vehicles: Real-time communication between vehicles and infrastructure (V2X) improves traffic flow and reduces accidents. Dynamic traffic management: AI-powered systems adjust signals and reroute vehicles to prevent congestion. Smart public transit: Connected busses and trains offer real-time tracking and optimized routes. Sustainable infrastructure smart grids: Advanced energy management systems dynamically balance supply and demand, integrate renewable energy, and reduce power outages. Energy-efficient buildings: Sensors monitor energy consumption and optimize heating, cooling, and lighting. Public safety-connected surveillance: AI-enabled cameras and sensors monitor public spaces, enhancing security and emergency response. Disaster management: Real-time alerts and resource deployment during natural disasters improve resilience. Environmental monitoring air quality sensors: IoT devices measure pollutants and provide actionable insights for reducing emissions. Smart waste management: Connected bins notify authorities when full, optimizing collection routes and reducing waste overflow. Enhanced citizen services e-governance platforms: High-speed connectivity allows seamless access to government services and information. Smart healthcare: Remote diagnostics, telemedicine, and wearable devices improve healthcare accessibility and outcomes.

5.3 Benefits of 5G and beyond for smart cities

Improved quality of life, personalized and efficient public services. Reduced commute times and improved air quality through smart transportation. Economic growth encourages innovation and entrepreneurship by providing robust digital infrastructure. Attracts investments through advanced technology ecosystems. Sustainability reduced carbon footprint through energy-efficient systems and optimized resource usage. Enhanced urban planning to support green initiatives.

5.4 Challenges in implementation

Infrastructure development has high costs of deploying 5G/B5G networks and upgrading legacy systems. Privacy and security manage data security and ensure compliance with privacy regulations in hyper-connected environments. Digital divide ensuring equitable access to smart city services across diverse socioeconomic groups. Interoperability standardizes technologies to enable seamless communication among devices from different manufacturers. The integration of 5G, B5G, and 6G technologies will redefine urban living by creating intelligent, sustainable, and citizen-centric cities. From autonomous transportation and green energy solutions to enhanced public safety, these advancements will transform how cities operate and evolve.

6. Security and privacy concerns

Challenges in securing 5G and IoT networks: Vulnerabilities in IoT devices, risks in network slicing, and virtualization. Emerging solutions: Blockchain, AI-based threat detection. As 5G and Beyond 5G (B5G/6G) networks enable unprecedented levels of connectivity and data exchange, security and privacy challenges become increasingly critical. These challenges arise from the complexity of these networks, the proliferation of IoT devices, and the integration of advanced technologies like AI and edge computing.

6.1 Key security concerns

Increased attack. Surface proliferation of IoT devices: Billions of interconnected devices create numerous entry points for cyberattacks. Decentralized networks: Distributed architectures, such as edge computing, increase the number of potential vulnerabilities. Advanced Persistent Threats (APTs): Sophisticated cyberattacks target critical infrastructure, such as smart grids and healthcare systems, potentially causing widespread disruption. Vulnerabilities in network slicing: Misconfigured or compromised network slices could expose sensitive data or disrupt critical services allocated to specific applications. Supply chain risks: Dependency on global supply chains for hardware and software increases the risk of malicious components or backdoors being embedded in network equipment. Quantum computing threats: Quantum computing could render traditional cryptographic methods obsolete, exposing data to decryption.

6.2 Key privacy concerns

Data overcollection in IoT devices and smart city infrastructure generates vast amounts of data, often collecting more information than necessary, raising privacy concerns. Location tracking is continuous monitoring of device locations that enables accurate tracking of individuals, potentially violating privacy rights. Lack of user control: Users often have limited visibility into how their data is collected, stored, and shared, leading to potential misuse. Data breaches: Sensitive information, including health records and financial data, stored or transmitted over these networks is at risk of being exposed.

6.3 Mitigation strategies

Strengthened security protocols zero-Trust Architecture (ZTA): Requires continuous verification of devices and users, even within the network. End-to-end encryption: Ensures data remains secure during transmission. AI-driven threat detection in AI and machine learning can identify and mitigate threats in real time by analyzing network behavior and anomalies. Advanced cryptography development of post-quantum cryptography to safeguard against quantum computing threats. Implementation of lightweight cryptography for resource-constrained IoT devices. Secure software development by encouraging secure coding practices and regular updates to minimize vulnerabilities in applications and firmware. Data minimization and anonymization collect only essential data and anonymize it to protect user identities and sensitive information. Regulatory compliance adherence to privacy regulations such as GDPR, CCPA, and other local laws to protect consumer rights. Clear policies on data collection, storage, and sharing. Secure device lifecycle management regularly updates and patches IoT devices. Implement robust mechanisms for decommissioning devices to prevent misuse of residual data.

6.4 Emerging technologies to enhance security and privacy

Blockchain provides decentralized and tamper-proof mechanisms for secure data exchange and identity management. Homomorphic encryption allows data to be processed while encrypted, eliminating the need to expose sensitive information during computation. Federated learning in AI training occurs locally on devices, ensuring sensitive data never leaves the device.

6.5 Challenges in addressing security and privacy

Balancing performance and security: High-speed, low-latency requirements may conflict with the computational demands of robust security protocols. Resource constraints: Many IoT devices have limited processing power and memory, making the implementation of advanced security measures difficult. Global collaboration: Coordinating security standards and practices across nations and industries is challenging but essential. User awareness: End users often lack awareness of security best practices, such as updating devices or recognizing phishing attempts. As 5G and B5G/6G networks continue to evolve, prioritizing security and privacy will be crucial for fostering trust and enabling widespread adoption. Governments, industry stakeholders, and researchers must collaborate to develop innovative solutions and standards that ensure a secure and privacy-preserving digital ecosystem.

7. Standards and regulations

Current 5G standards for IoT and smart cities. Anticipated standards for B5G/6G. Policy and regulatory challenges in wireless communication advancements. The deployment and operation of 5G, Beyond 5G (B5G/6G), and IoT technologies require robust standards and regulatory frameworks to ensure interoperability, security, fairness, and efficient spectrum utilization. These frameworks guide the development, implementation, and governance of advanced communication networks worldwide.

7.1 Key objectives of standards and regulations

Interoperability ensures seamless communication between devices, networks, and systems from different manufacturers and operators. Security is protecting networks, data, and devices from cyber threats and privacy violations. Spectrum management optimizes the allocation and use of limited-frequency resources. Fair access prevents monopolization and ensures equitable access to technologies globally. Innovation enablement fosters innovation while balancing safety, privacy, and sustainability concerns.

7.2 Major organizations defining standards

The International Telecommunication Union (ITU) governs global spectrum allocation and international standards for telecommunication. Develops frameworks for International Mobile Telecommunications (IMT), including IMT-2020 (5G) and future IMT-2030 (6G). The 3rd Generation Partnership Project (3GPP) defines technical specifications for 5G, including Release 15: Initial 5G standardization, Release 16 and beyond: Advanced capabilities for ultra-reliable low-latency communication (URLLC), massive IoT, and vehicular communication. The Institute of Electrical and Electronics Engineers (IEEE) develops standards for wireless communication, including Wi-Fi (802.11 series), which complements cellular networks in IoT environments. The European Telecommunications Standards Institute (ETSI) leads efforts in network function virtualization (NFV) and multi-access edge computing (MEC) to enhance 5G and B5G capabilities. The Internet Engineering Task Force (IETF) focuses on developing protocols for the Internet, ensuring secure and efficient communication in IoT and other applications.

7.3 Key regulatory bodies

Federal Communications Commission (FCC)—USA regulates spectrum allocation and ensures compliance with communication standards in the United States. European Union (EU) implements GDPR for data privacy and develops regulations to harmonize 5G deployment across member states. Ministry of Industry and Information Technology (MIIT)—China oversees spectrum allocation and promotes domestic 5G and B5G innovations. The Telecom Regulatory Authority of India (TRAI) sets policies for spectrum pricing, 5G deployment, and IoT adoption in India. International regulators collaborate through organizations like the ITU to create harmonized policies and frameworks for global interoperability.

7.4 Key areas of regulation

Spectrum allocation regulations determine frequency bands for 5G, such as low-band (<1 GHz): For wide coverage, mid-band (1–6 GHz): Balancing coverage and speed, and high-band (>24 GHz, mmWave): For ultra-high speeds and dense urban areas. Emerging B5G/6G networks are exploring sub-terahertz and terahertz frequencies. Security and privacy data protection laws: GDPR (EU), CCPA (USA), and similar laws govern the collection, storage, and usage of data. Cybersecurity standards: Guidelines to mitigate threats in critical infrastructure, such as IoT and smart cities. IoT-specific regulations. Device certification: Ensuring IoT devices meet security and performance standards. Network congestion management: Preventing overload

DOI: http://dx.doi.org/10.5772/intechopen.1009925

in densely populated IoT networks. Environmental sustainability energy efficiency standards: Encourage the development of energy-efficient hardware and protocols. E-waste regulations: Manage the disposal and recycling of outdated devices.

7.5 Challenges in standards and regulations

Global coordination: Diverging national interests complicates the creation of unified global standards. Rapid technological evolution: Regulatory frameworks struggle to keep pace with advancements in 5G, B5G/6G, and IoT. Spectrum scarcity: Balancing spectrum allocation between commercial, military, and public safety applications is challenging. Ensuring security and privacy by defining globally enforceable cybersecurity and data privacy standards is complex due to jurisdictional differences. 6G standardization: Early discussions are focused on enabling sub-terahertz communication, quantum security, and AI-driven networks. Stronger security frameworks: Emphasis on post-quantum cryptography, secure device lifecycles, and AI-based threat detection. Sustainability: Enhanced regulations for energy-efficient networks and environmentally friendly deployment practices.

8. Future directions and challenges

Open research challenges in wireless communication for IoT and smart cities: Energy efficiency, scalability, and interoperability; vision for 6G and beyond; the fully connected world; societal impacts of advancements in wireless communication. As 5G networks are deployed and 6G begins to take shape, the future of connectivity and IoT presents exciting opportunities and complex challenges. The next few years will be defined by rapid technological evolution, deep integration of AI, edge computing, and quantum technologies, and the transformation of industries like healthcare, transportation, energy, and urban planning. Here is a closer look at future directions and the challenges that will shape the evolution of connectivity and smart technologies.

8.1 Future directions

6G networks and beyond terahertz communication: 6G will explore the use of sub-terahertz and terahertz frequencies (100 GHz–1 THz), offering data rates that could exceed 1 Tbps and support instantaneous data transfer. This will enable real-time holographic communications, immersive virtual environments, and other data-intensive applications. AI-driven networks: 6G will incorporate AI at every layer of the network—from resource management to self-healing networks and predictive maintenance. AI will enable networks to adapt dynamically based on demand, user behavior, and environmental factors. Quantum communications: Quantum computing and quantum encryption could provide unbreakable security and dramatically enhance privacy. Quantum technologies will also play a role in overcoming the limitations of classical cryptography. Holographic and immersive communication: 6G will likely enable holographic communication, offering immersive, real-time 3D experiences for virtual meetings, entertainment, and education, paving the way for a "tactile internet" that allows touch and feel sensations over the internet. Smart cities and autonomous systems ubiquitous smart cities: With advanced 5G and 6G, cities will become fully interconnected ecosystems. Smart cities will feature AI-driven traffic management, predictive maintenance, energy optimization, real-time pollution monitoring, and autonomous public services (e.g., waste collection, healthcare, and transportation). Autonomous transportation: The transition toward self-driving cars, autonomous drones, and driverless public transport will be accelerated by low-latency, high-reliability 5G/6G networks. Vehicle-to-everything (V2X) communication will create an interconnected transportation ecosystem, reducing accidents and traffic congestion. Edge computing: Edge networks will become more prevalent, reducing latency by bringing computing power closer to the end devices. This will support autonomous systems, real-time AI decision-making, and industrial IoT applications. IoT and Industry 4.0 industrial IoT (IIoT): Advanced networks will revolutionize industries by enabling smart factories, predictive maintenance, and real-time monitoring. Industrial machines will communicate seamlessly, optimizing production and supply chains. Healthcare revolution: Remote patient monitoring, telemedicine, AI-powered diagnostics, and robotic surgeries will be enabled by high-speed, low-latency connectivity. Wearable devices will monitor vital signs and send data to healthcare providers in real time, transforming personalized healthcare. Smart agriculture: IoT sensors and AI algorithms will enable precision farming, optimizing irrigation, soil quality, pest control, and crop yields. Autonomous farming machinery will improve efficiency. Sustainability and green tech energy-efficient networks: The demand for energy-efficient communication networks will increase, driving the development of green technologies such as low-power IoT devices, energy harvesting, and green data centers. Sustainable IoT: The focus on minimizing the environmental impact of the growing IoT ecosystem will lead to the development of energy-efficient protocols and low-power IoT standards.

8.2 Challenges to overcome

Network scalability: With the explosion of connected devices and increased demand for high-speed data, scaling networks efficiently remains a critical challenge. 5G and 6G networks will need to support millions of IoT devices, millimeter-wave communication, and real-time analytics, all while ensuring optimal performance. The development of network slicing—dedicated virtual networks for specific applications—will help, but managing such vast and complex networks will require advanced orchestration and AI-based automation. Security and privacy increased attack surface: The interconnection of billions of devices in IoT and smart cities creates numerous vulnerabilities. Networks will be susceptible to cyberattacks, data breaches, and privacy violations unless strong measures like AI-powered threat detection, quantum encryption, and zero-trust security models are implemented. Data privacy: With the proliferation of smart devices collecting personal data, ensuring data privacy is a major concern. Governments and organizations must enforce stricter regulations to guarantee that data is protected and user consent is obtained. Spectrum management: As the demand for high-frequency bands, such as millimeter-wave and terahertz, increases, spectrum congestion may become a limiting factor. Efficient spectrum sharing and dynamic spectrum management technologies will be critical in addressing these challenges. Regulatory challenges global coordination: Different countries have different regulatory frameworks for 5G/6G deployment, which could slow down international standardization efforts and prevent seamless interoperability across borders. Data sovereignty: As data becomes increasingly globalized, countries will need to address data sovereignty issues, ensuring that data stored and processed across borders complies with local laws and regulations. Technological disparity: There is a risk that emerging technologies such as 5G and 6G could exacerbate the digital divide, leaving rural areas and developing countries with limited access to advanced networks. Efforts must be made to bridge the gap and ensure equitable access to digital technologies. Environmental Impact Energy consumption: The implementation of 5G/6G infrastructure, such as base stations and edge devices, will significantly increase energy consumption. While there is a drive for energy-efficient solutions, sustainable deployment practices need to be a priority to minimize the environmental impact. The future of 5G, B5G, and IoT holds enormous promise, with transformative applications across industries. However, to fully realize this potential, we need to overcome several technical, regulatory, and societal challenges. The next generation of connectivity will depend on balancing innovation with security, sustainability, and equitable access.

9. Conclusion

Potential impact on IoT and smart city development and closing thoughts on the future of wireless communication. The advent of 5G and the forthcoming Beyond 5G (B5G)/6G technologies represent transformative milestones in the world of wireless communications, with profound implications for the Internet of Things (IoT) and the development of smart cities. These advancements bring about unprecedented opportunities for creating hyper-connected, intelligent environments that are more efficient, sustainable, and responsive to the needs of urban populations. As we move beyond the 5G era, the capabilities of these next-generation networks will enable ultra-high-speed communications, ultra-low latency, and massive connectivity. These innovations will power the next wave of autonomous vehicles, AI-driven smart cities, industrial automation, and advanced healthcare solutions. The deep integration of IoT, AI, and edge computing will empower cities to optimize traffic flow, manage energy resources efficiently, and ensure public safety through real-time data and automation.

However, as exciting as these advancements are, there are significant challenges ahead. Issues related to security, privacy, spectrum management, regulatory compliance, and scalability will require careful attention. The sheer scale of device interconnectivity, the growing demands for data privacy, and the need for regulatory harmonization across nations are all hurdles that must be overcome for these technologies to reach their full potential. Ultimately, the realization of truly smart cities and the widespread adoption of IoT will depend on a balanced approach where technological innovation is paired with robust standards, regulatory frameworks, and a focus on sustainability and security. The collaboration between governments, industries, and researchers will be crucial in driving the next-generation infrastructure that can support this vision. As we look toward the future, 6G and its successors promise even greater advancements, offering the potential for instantaneous communication, holographic interactions, and AI-driven ecosystems that blur the line between the physical and digital worlds. These technologies will undoubtedly reshape our cities, economies, and the very way we interact with the world around us. In conclusion, the path ahead is filled with immense possibilities but also substantial challenges. By continuing to innovate, collaborate, and address key concerns, we can unlock the full potential of 5G, B5G, and 6G to build smarter, safer, and more sustainable urban environments for future generations.

Author details

Srinivas Ramavath*, Umesh Chandra Samal, Prasanta Kumar Patra, Pattepu Sunil and Bhargav Appasani
School of Electronics Engineering, Kalinga Institute of Industrial Technology (KIIT) Deemed to be University, Bhubaneswar, India

*Address all correspondence to: srinivas2012iitg@gmail.com

References

[1] Shehab MJ, Kassem I, Kutty AA, Kucukvar M, Onat N, Khattab T. 5G networks towards smart and sustainable cities: A review of recent developments, applications and future perspectives. IEEE Access. 2021;**10**:2987-3006

[2] Minoli D, Occhiogrosso B. Practical aspects for the integration of 5G networks and IoT applications in smart cities environments. Wireless Communications and Mobile Computing. 2019;**2019**(1):5710834

[3] Khanh QV, Hoai NV, Manh LD, Le AN, Jeon G. Wireless communication technologies for IoT in 5G: Vision, applications, and challenges. Wireless Communications and Mobile Computing. 2022;**2022**(1):3229294

[4] Milovanovic D, Pantovic V, Bojkovic N, Bojkovic Z. Advanced human centric 5G-IoT in a smart city: Requirements and challenges. In: Human Centered Computing: 5th International Conference, HCC 2019, Čačak, Serbia, August 5-7, 2019, Revised Selected Papers 5. Springer International Publishing; 2019. pp. 285-296

[5] Rafique W, Barai J, Fapojuwo AO, Krishnamurthy D. A survey on beyond 5g network slicing for smart cities applications. IEEE Communications Surveys & Tutorials. 2024

[6] Chettri L, Bera R. A comprehensive survey on Internet of Things (IoT) toward 5G wireless systems. IEEE Internet of Things Journal. 2019;**7**(1):16-32

[7] da Silva MM, Guerreiro J. On the 5G and beyond. Applied Sciences. 2020;**10**(20):7091

[8] Guevara L, Cheein FA. The role of 5G technologies: Challenges in smart cities and intelligent transportation systems. Sustainability. 2020;**12**(16):6469

[9] Patra MK, Sahoo S, Sahoo B, Turuk AK. Cloud computing for a secure smart city beyond 5G. In: Secure and Intelligent IoT-Enabled Smart Cities. IGI Global; 2024. pp. 91-116

[10] Ogbodo EU, Abu-Mahfouz AM, Kurien AM. A survey on 5G and LPWAN-IoT for improved smart cities and remote area applications: From the aspect of architecture and security. Sensors. 2022;**22**(16):6313

[11] Alraih S, Shayea I, Behjati M, Nordin R, Abdullah NF, Abu-Samah A, et al. Revolution or evolution? Technical requirements and considerations towards 6G mobile communications. Sensors. 2022;**22**(3):762

[12] Tachikawa K. A perspective on the evolution of mobile communications. IEEE Communications Magazine. 2003;**41**(10):66-73

[13] Kumar S, Zahn C. Mobile communications: Evolution and impact on business operations. Technovation. 2003;**23**(6):515-520

[14] Ramavath S, Patra PK, Samal UC. Evaluation and analysis of prototype filters for 5G FBMC system. In: 2023 IEEE 20th India Council International Conference (INDICON), Hyderabad, India. 2023. pp. 503-508. DOI: 10.1109/INDICON59947.2023.10440678

[15] Ramavath S, Akhil R, Swayamsiddha S, Samal UC. Energy incentive for packet relay in internet of things networks using 5G NR. In: 2023

IEEE 3rd International Conference on Applied Electromagnetics, Signal Processing, & Communication (AESPC), Bhubaneswar, India. 2023. pp. 1-5. DOI: 10.1109/AESPC59761.2023.10390256

[16] Ezema ME, Okoye FA, Okwori AO. A framework of 5G networks as the foundation for IoTs technology for improved future network. International Journal of Physical Sciences. 2019;**14**(10):97-107

[17] Song H, Srinivasan R, Sookoor T, Jeschke S. Smart Cities: Foundations, Principles, and Applications. John Wiley & Sons; 2017

[18] Skouby KE, Lynggaard P. Smart home and smart city solutions enabled by 5G, IoT, AAI and CoT services. In: 2014 International Conference on Contemporary Computing and Informatics (IC3I). IEEE; 2014. pp. 874-878

Section 3

Antenna Design and Measurement for 5G/6G Systems

Chapter 4

Automatized Measurement Setup to Determine 5G and 6G Radiation Pattern for Miniaturized Arrays Antennas

Vanessa Przybylski Ribeiro Magri, Leni Joaquim de Matos, Pedro Vladimir Gonzalez Castellanos, Vítor Luiz Gomes Mota and Rodrigo Amitrano Bilobran

Abstract

A swivel low-cost prototype is proposed for measuring the radiation pattern of millimeter wave antennas and antenna arrays applied to mobile services such as 5G and 6G and other wireless systems like RFID. Furthermore, the proposed measurement setup uses a pair of identical antennas to obtain the radiation pattern. The prototype uses a stepper motor controlled by Arduino Uno, an open-source software that allows a complete 360° rotation. The measurement setup utilizes a swivel model, where the transmitter antenna is connected to a signal generator, and the receiver antenna is connected to a signal analyzer, both controlled via a GPIB-USB interface. LabVIEW is employed to manage the equipment and measure the received power levels. Additionally, MATLAB, integrated into the same LabVIEW VI, plots the radiation pattern in rectangular and polar formats. A second calibration approach for verifying the radiation pattern diagram of a commercial directional antenna is also presented. The simulated and measured results are compared and validated in real-world environments and anechoic chamber.

Keywords: antenna, radiation pattern, low-cost prototype, swivel, Arduino uno

1. Introduction

Wireless communication systems are rapidly expanding worldwide, driven by the increasing number of Internet users on smart devices and smartphones. The fifth generation (5G) has emerged to handle a hundred times the current traffic and deliver speeds ten times faster than the 4G generation [1–4].

A key component of these systems is the antenna, which is responsible for transmitting and receiving electromagnetic waves. The requirements for a 5G antenna include compactness, low profile and cost, high directivity, and ease of fabrication [5–7]. By analyzing the radiation pattern, it is possible to evaluate whether the antenna meets these

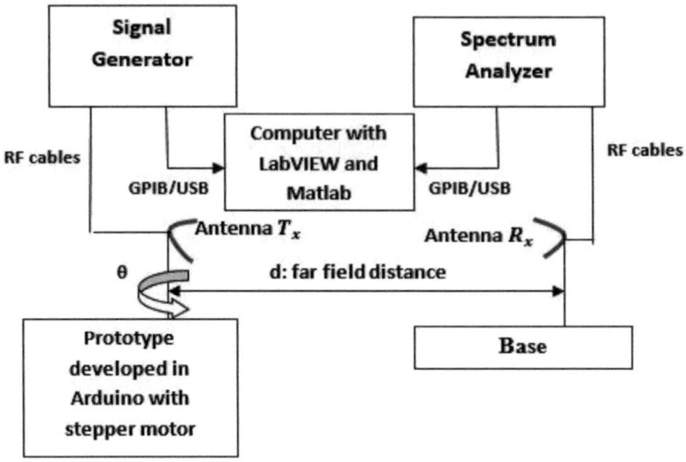

Figure 1.
Swivel prototype and automatized measurement setup proposed.

requirements. Additionally, parameters such as gain, bandwidth, impedance, half-power beamwidth (HPBW), and first null beamwidth (FNBW) [8] can also be determined.

The technique used to estimate the radiation pattern with high precision, an anechoic chamber, and a reference antenna are typically required [9, 10]. However, constructing such an environment can be prohibitively expensive for some research centers. As a result, several authors have sought to develop low-cost measurement setups.

The measurement setup proposed in [11, 12] is specifically tailored for Yagi antennas. In [13, 14], the authors introduce a system that uses a probe to connect the antenna under test (AUT). A measurement system based on an field programmable gate array (FPGA) is described in [15]. Additionally, [16] presents a low-cost setup implemented with Arduino, incorporating a reference antenna.

This chapter aims to present the research and development of a cost-effective alternative method for characterizing antennas through their radiation patterns. The proposed approach involves an Arduino-based prototype that enables antenna rotation and utilizes a pair of identical antennas, thereby eliminating the need for a dedicated reference antenna, as shown in **Figure 1**. Another setup is also proposed when a reference antenna is available, and the goal is to obtain the radiation pattern diagram of another antenna.

2. Proposed system

The platform uses a hybrid stepper motor to enable antenna rotation. The commercial model 28BYJ-48 was selected due to its good angle resolution (0.088°), combining precision and torque, which are characteristic features of this type of motor [17]. This motor connects to the ULN 2003 driver, which supports currents up to 500 mA and operates at 1 V. This platform meets several technical requirements, including USB connectivity, digital and analog inputs/outputs, the possibility of external power supply, and the option to change the microcontroller, in addition to its low cost for project implementation [18].

To enhance the user experience, an IR (infrared) module and an LCD display are incorporated into the project. The IR module controls the motor's rotation,

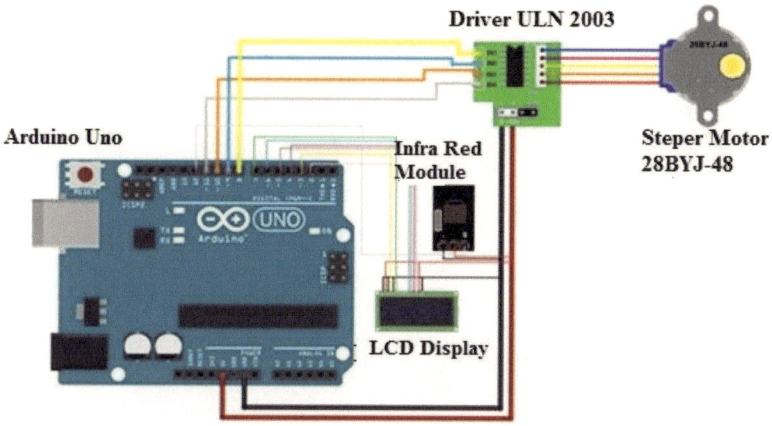

Figure 2.
Arduino Uno controlling stepper motor.

so when the user presses the right or left button on the remote control, the stepper motor moves accordingly. The LCD display shows the device's current angle. **Figure 2** illustrates all the devices simultaneously controlled by the Arduino Uno.

After the logical connection between the devices and the Arduino Uno platform is established, the definition of the rotation variable is necessary. It represents the desired angle step for the stepper motor, that is, the number of degrees it will rotate. In this particular case, it adopted a rotation increment of 10 degrees, as this interval allows for the collection of sufficient samples to be used in the second part of the project, which involves measuring the power received per angle.

This variable is in the function "myStepper.step" in Arduino Uno library. This function does not directly accept the desired degree as a parameter, as its predefined parameter is the number of steps. To enable this function to understand and execute the desired rotation, a simple proportion must be applied.

This motor model is capable of completing one full rotation in 2048 steps; therefore, 360° correspond to 2048 steps. Thus, the proportion used can be represented by the following equation:

$$N = \frac{(2048) * (\alpha)}{360} \tag{1}$$

where: N corresponds of to the number of steps and (α) the desired angle.

Insulating materials were selected for the development of the prototype structure shown in **Figure 3**. Wood is used as the support for securing the AUT, while polyvinyl chloride (PVC) is employed to connect the top and bottom sections, providing the necessary height for the AUT. As designed, the prototype is portable, and an medium-density fiberboard (MDF) enclosure is fabricated to house the wiring, electrical connections, and other components, protecting them from user interference.

The swivel prototype described, along with the Anritsu MS3700A signal generator (SG), forms part of the transmission system. The receiver system uses Anritsu MS2692A spectrum analyzer (SA) to acquire the signal. LabVIEW software controls both pieces of equipment, enabling automation of the measurement setup.

Although an ethernet interface could be used to control the system, the equipment is controlled via the GPIB interface, as shown in **Figure 4**. The primary reasons for choosing GPIB over ethernet are safety, the ability to control two or more devices simultaneously, and the capability to operate without an Internet connection [19].

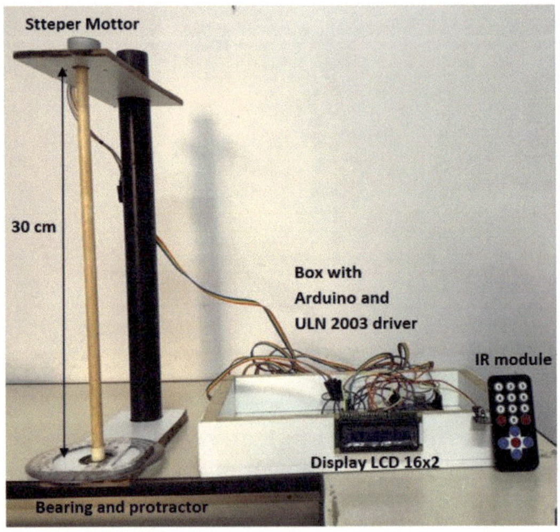

Figure 3.
Swivel prototype proposed controlled by Arduino Uno.

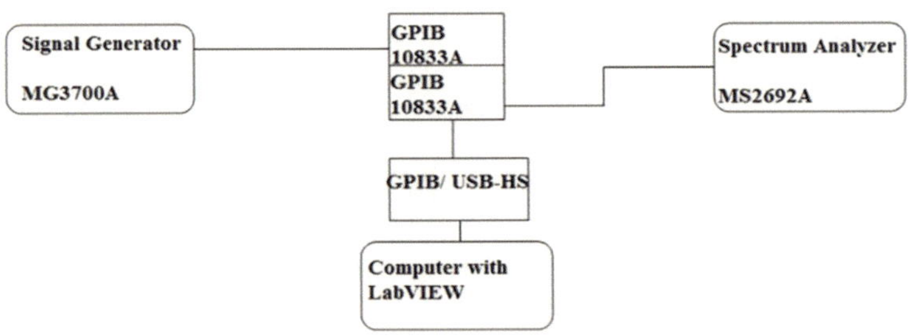

Figure 4.
Automatized equipment control by LabVIEW.

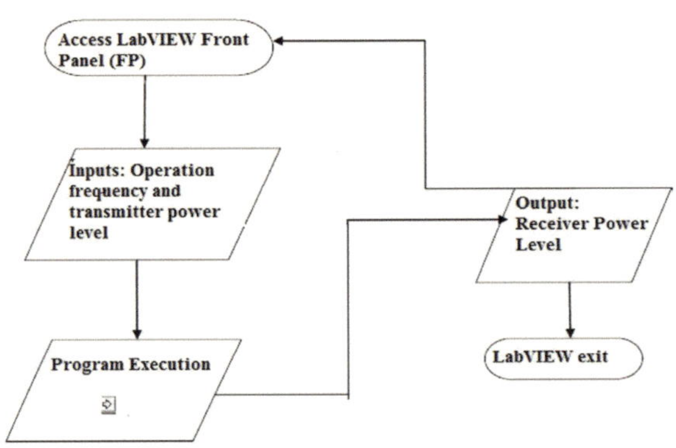

Figure 5.
Calibration program flowchart.

DOI: http://dx.doi.org/10.5772/intechopen.1009431

A calibration program was developed in LabVIEW with the goal of configuring the signal generator (SG) and spectrum analyzer (SA). The loss introduced by the RF cables is measured through a back-to-back connection, allowing the receiver power to

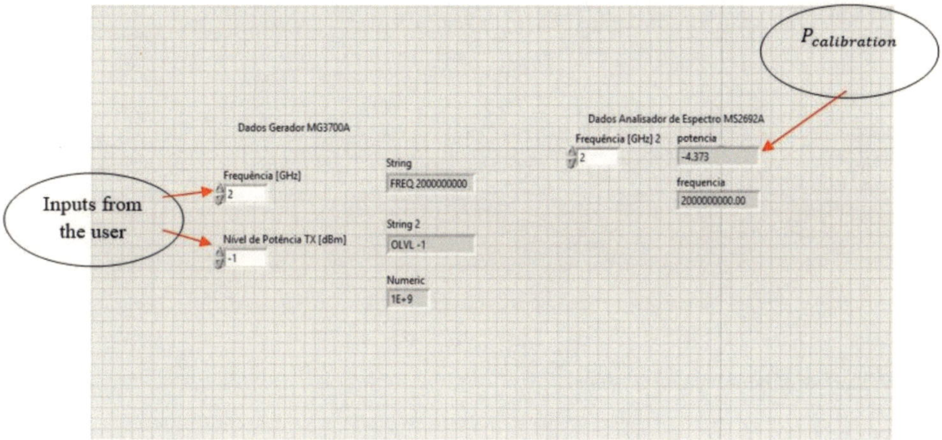

Figure 6.
LabVIEW interface of calibration program.

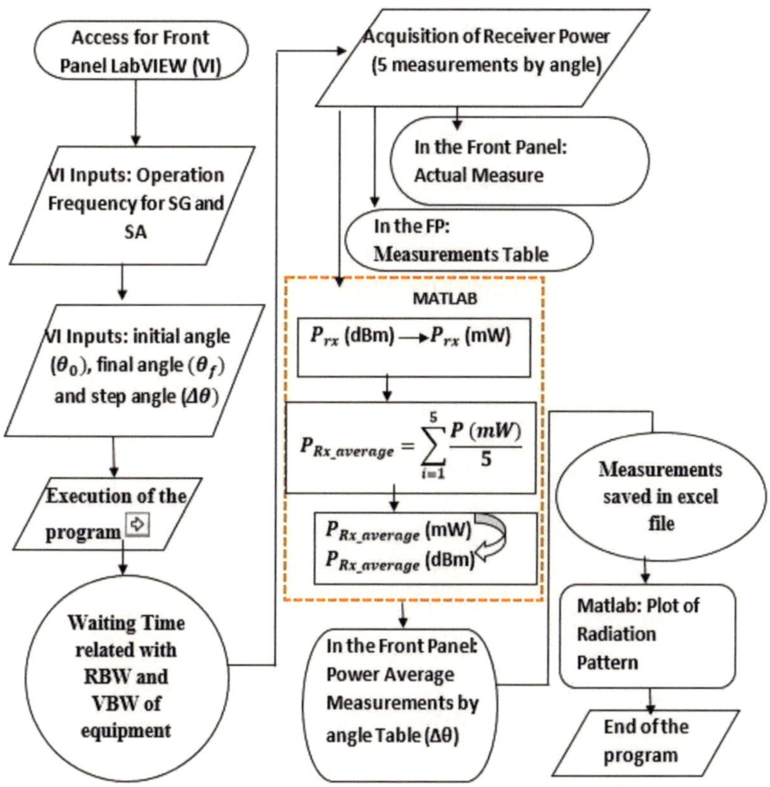

Figure 7.
Measurement program flowchart.

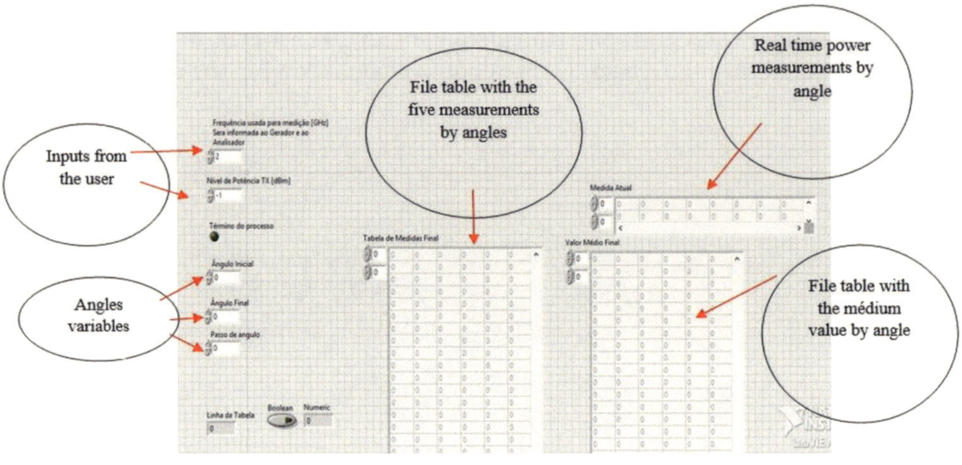

Figure 8.
LabVIEW interface of measurement program.

be determined. This enables the calibration of the system. The operating frequency of the SG and SA, along with the transmitter power from the SG, serves as inputs, while the receiver power is the output of the program, as shown in the flowchart in **Figure 5** and the LabVIEW interface present in **Figure 6**.

A measurement program was developed in LabVIEW based on the flowchart shown in **Figure 5**. The objective is to determine the angle variation and acquire the measurements needed to construct the radiation pattern. The program has five inputs: the operating frequency for the SG and SA, the transmitter power level from the SG, the initial angle, the stop angle, and the angle step as seen in **Figure 7**.

In this specific application, an angle step of 10° is used to generate the radiation pattern with reasonable precision; however, this step size can be reduced to obtain a more accurate radiation pattern. The output of the program is the receiver power level, with five measurements taken for each angle to minimize instantaneous errors caused by environmental factors, thus increasing the reliability of the results. In **Figure 8**, the interface of the LabVIEW software with the measurement program can be observed.

3. Measurement setup and antennas

The proposed system can operate with various types of antennas to determine their radiation pattern. This section demonstrates the validation of this system with two different types of antennas: a prototype antenna arrays for the 3.5 and 4.8 GHz bands and a commercial wideband directional printed antenna (1.4 to 10.5 GHz).

3.1 Prototype antenna arrays

The measurement setup consists of the swivel prototype, a computer running LabVIEW to automate the control of the equipment, and a pair of miniaturized antennas or an array for characterization.

Six pairs of arrays, combining rectangular and circular elements, were developed, simulated, and fabricated to demonstrate the system's efficacy. Those arrays are designed to operate at 3.5 and 4.8 GHz, frequencies under consideration for

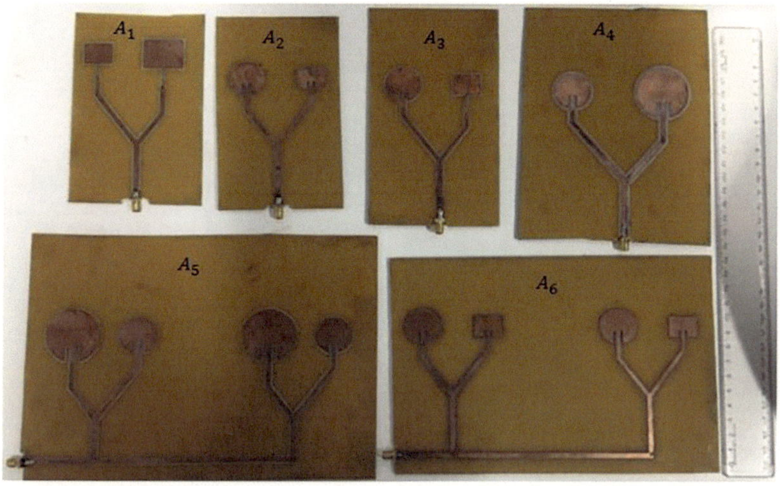

Figure 9.
Arrays for 5G and RFID services under tests.

5G networks in Brazil, China, South Korea, Japan, the United States, and the European Union [20], as well as at 2.4 GHz, a frequency used for Radio Frequency Identification (RFID) in Brazil. **Figure 9** shows such arrays.

One array is connected to the swivel prototype and the signal generator to form the transmitter system, serving as the antenna under test (AUT). The other array is connected to the spectrum analyzer to complete the receiver system, as shown in **Figure 10**, where parameters d and h represent the minimum distance between the antennas and between the elements and the floor, respectively, to define the far-field region.

For miniaturized antennas and arrays, this distance corresponds to 10 times the wavelength of the lowest frequency [21]. **Table 1** shows the minimum distance for each array.

Due to the reciprocity theorem and the fact that two identical arrays produce the same radiation pattern [22], there is no need for a reference antenna in this measurement setup, as is typically required in other setups [23, 24].

To generate the radiation pattern, a code developed in MATLAB® calculates the average measurements at each angular position. The code accesses the measurement spreadsheet, converts the power average values in dBm units and normalize. Then, the rectangular and polar plots of the measurement are generated based on the normalized values, using the *plot* and *polar* functions, respectively.

In this same code, a section for generating the simulated radiation pattern curve is included to enable a better comparative evaluation between the results obtained through simulation and measurement. Then, the simulated values, exported from the HFSS simulation, are plotted alongside the measured results.

3.2 Commercial wideband directional antenna

This section demonstrates a variation of the methodology presented to determine the radiation pattern of an antenna using a reference antenna BTA 118 with known HPBW and gain. For transmission, a signal generator coupled to a properly characterized RF cable and a reference antenna were used to determine the radiation pattern

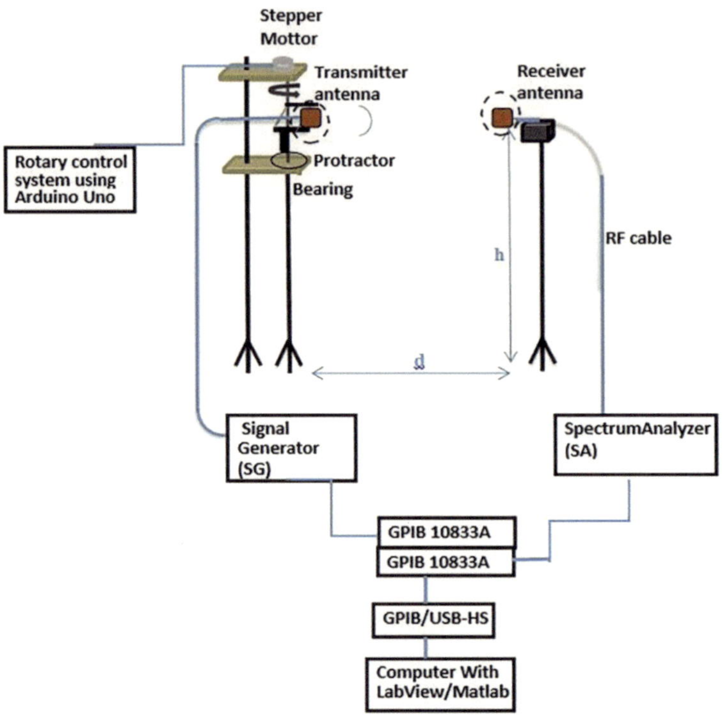

Figure 10.
Radiation pattern measurement setup indoor environments using a pair of identical printed antennas under test.

Parameter	A_1	A_2	A_3	A_4	A_5	A_6
Far-field zone (cm)	117.60	83.30	85.71	120.00	125.00	85.71

Table 1.
Minimum distances for each array.

of the directional antenna operating from 1.4 to 10.5 GHz. For reception, the test antenna was positioned in the direction of the transmitting antenna with equivalent heights, respecting maximum radiation in both polarizations.

The test antenna, whose radiation pattern is to be measured, must be coupled to a stepper motor for precise rotation in the horizontal plane. Then, the received power will be used in the calculation of the radiation pattern. In this setup, for signal generation of continuous wave signal, the Anritsu MG3700A signal generator was used, along with an RG213 cable and a 10 dBi gain horn antenna with a 60° HPBW and a frequency range of 1 to 18 GHz. In the reception part, the antenna used for testing where the radiation pattern must be measured was fixed on an axis coupled to a Nema 17 stepper motor, and an SMA/N RF RG402 cable was used for connection to the RSA306B spectrum analyzer. **Table 2** presents the equipment used in the measurement, and **Figure 11** shows the measurement setup.

The setup shown in **Figure 11** was in an environment free of reflections (open área) that could affect the measured signal level. Additionally, the use of minimum distance between the TX and RX antennas avoids any reflected components from the ground and respecting the Fraunhofer distance.

Item	Specification
Horn Antenna BT 118 (1 to 18 GHz)	10 dBi
Antenna R_x (2 to 10.5 GHz)	7 dBi
T_x Power	3 dBm
T_x RG213 cable loss	1.7 dB
R_x RG402 cable loss	1.4 dB

Table 2.
Setup equipment and their specifications.

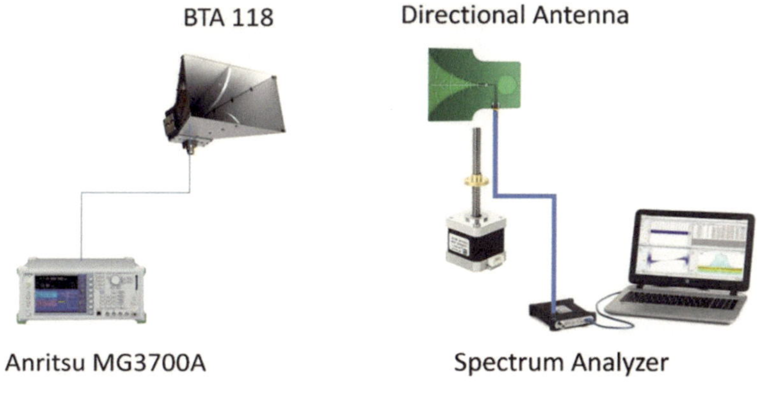

Figure 11.
Radiation pattern measurement setup in outdoor environments using a horn referential antenna and a directional printed antenna under test.

$$d_{min} > \frac{2D^2}{\lambda} \quad (2)$$

The antennas are positioned at the same height, approximately 1.5 meters above the ground, and in the same polarization. Before starting the radiation pattern calculation process, a reference measurement must be performed with the antennas pointed at each other in the direction of maximum radiation, in order to verify if the propagation loss obtained is close to the theoretical loss, calculated using the Friis model [8]. After the reference measurement, the radiation pattern measurement of the antenna can begin. To determine the radiation pattern, a narrowband signal (continuous wave) was emitted by the TX antenna. The RX antenna was rotated from 0 to 180° in steps of 10° in the horizontal plane, collecting 50 signal samples at each angle and saving the obtained data for later processing of the received power.

4. Results

This section presents the results of the radiation patterns obtained for the commercial directional wideband antenna and for the manufactured antenna array.

4.1 Prototype antenna arrays

Firstly, the measurement setup was in a real-world environment, specifically a laboratory room, as shown in **Figure 12**. Absorptive materials are utilized to minimize reflections and other factors that could compromise the accuracy of the results. A_1, A_6, A_3, A_2 are tested in 3.4 and 4.8 GHz, 5G frequencies, respectively. **Figures 13–16** present the normalized rectangular and polar radiation pattern measured (red line) and simulated (blue dashed line) for these arrays.

A_4 and A_5 are tested at RFID frequency 2.4 GHz. **Figures 16** and **17** present the rectangular and polar radiation pattern measured (line in red) and simulated (dashed line in blue) for those arrays (**Figure 18**).

Upon analyzing **Figures 13–17**, it is evident that the simulated and measured results are similar, meaning both curves exhibit the same behavior. The peaks and valleys are aligned at the same angles, indicating that the prototype and automated measurement setup are efficient, low-cost methods suitable for real-world scenarios. Despite the similarities, some differences are observed. This can be attributed to the measurements being conducted in a real-world environment, subject to multipath effects and various reflections.

In the second phase, measurements were conducted in an anechoic chamber, the ideal environment for measuring radiation patterns. Arrays A_4 and A_6 were tested, and the results obtained in the anechoic chamber (black dashed and dotted line) were compared with the simulation results (blue dashed line) and the real-world environment (red line), as shown in **Figures 19** and **20**.

When comparing these curves across the different scenarios, they exhibit similar behavior. Some discrepancies are observed in the real-world scenario curve, as compared to the others, due to the non-ideal conditions of the environment. However, given the similarities between the curves, it can be concluded that the method using a pair of identical antennas and the low-cost prototype is effective.

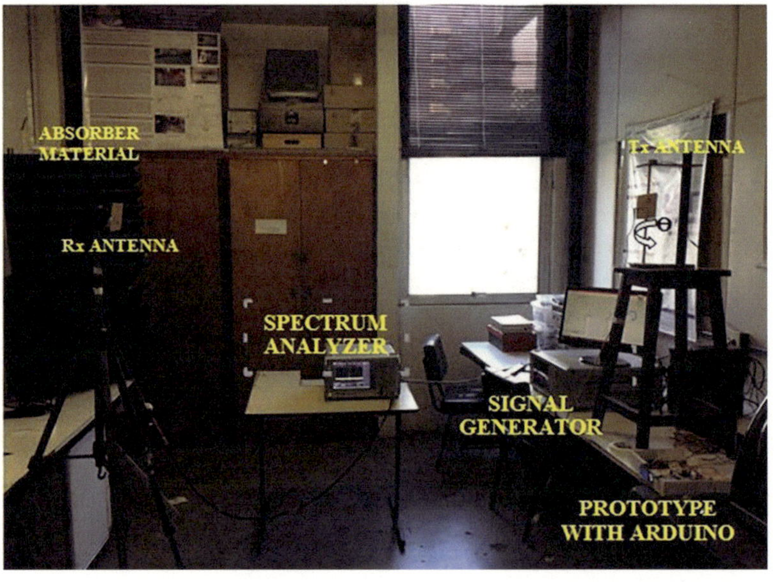

Figure 12.
Measurement setup applied in a real environment.

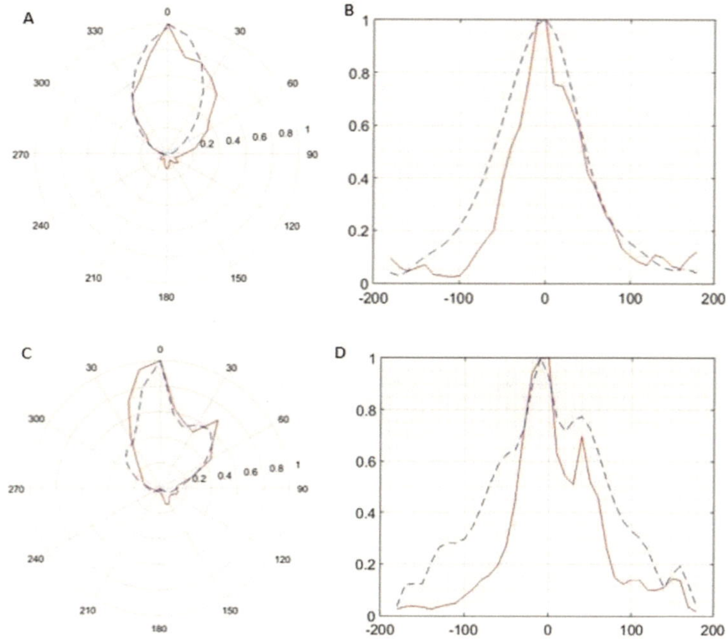

Figure 13.
Radiation pattern for A_1 array operating in 3.4 GHz: (a) rectangular for $\phi = 0^{\circ}$; (b) polar for $\phi = 0^{\circ}$; (c) rectangular for $\phi = 90^{\circ}$; (d) rectangular for $\phi = 90^{\circ}$.

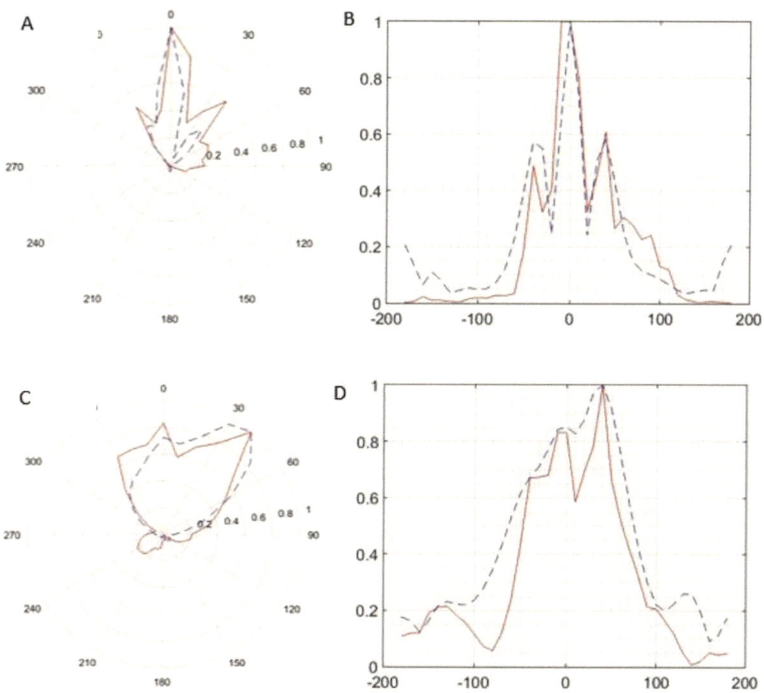

Figure 14.
Radiation pattern for A_6 array operating in 3.4 GHz: (a) rectangular for $\phi = 0^{\circ}$; (b) polar for $\phi = 0^{\circ}$; (c) rectangular for $\phi = 90^{\circ}$; (d) rectangular for $\phi = 90^{\circ}$.

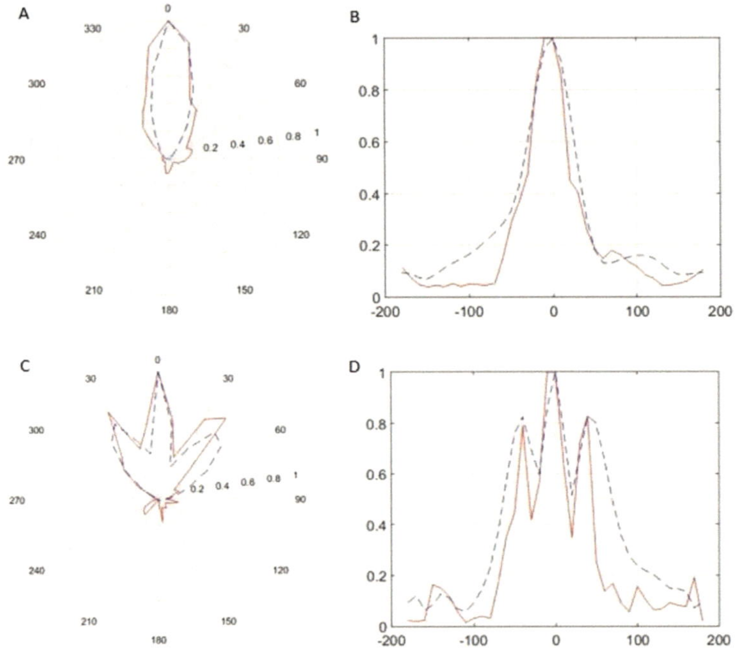

Figure 15.
Radiation pattern for A_2 array operating in 3.6 GHz: (a) rectangular for $\phi = 0^\circ$; (b) polar for $\phi = 0^\circ$; (c) rectangular for $\phi = 90^\circ$; (d) rectangular for $\phi = 90^\circ$.

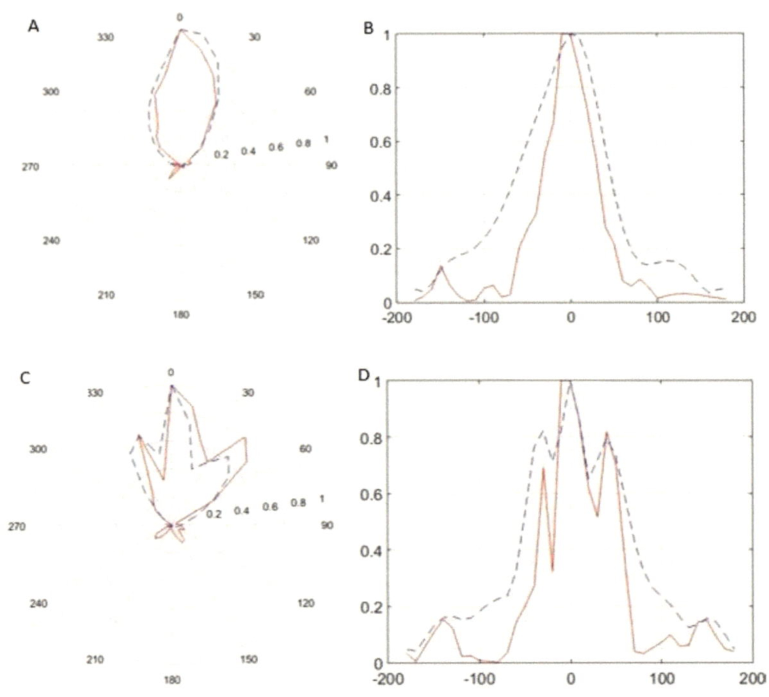

Figure 16.
Radiation pattern for A_3 array operating in 4.8 GHz: (a) rectangular for $\phi = 0^\circ$; (b) polar for $\phi = 0^\circ$; (c) rectangular for $\phi = 90^\circ$; (d) rectangular for $\phi = 90^\circ$.

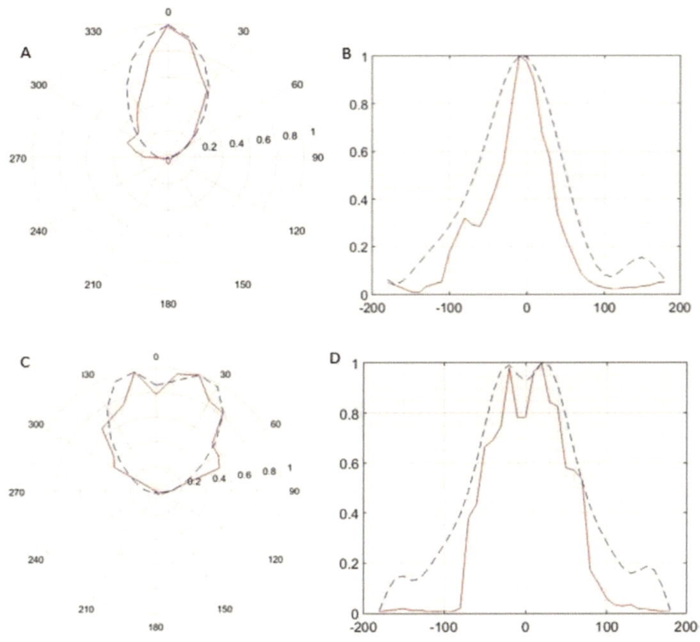

Figure 17.
Radiation pattern for A_4 array operating in 2.4 GHz: (a) rectangular for $\phi = 0^\circ$; (b) polar for $\phi = 0^\circ$; (c) rectangular for $\phi = 90^\circ$; (d) rectangular for $\phi = 90^\circ$.

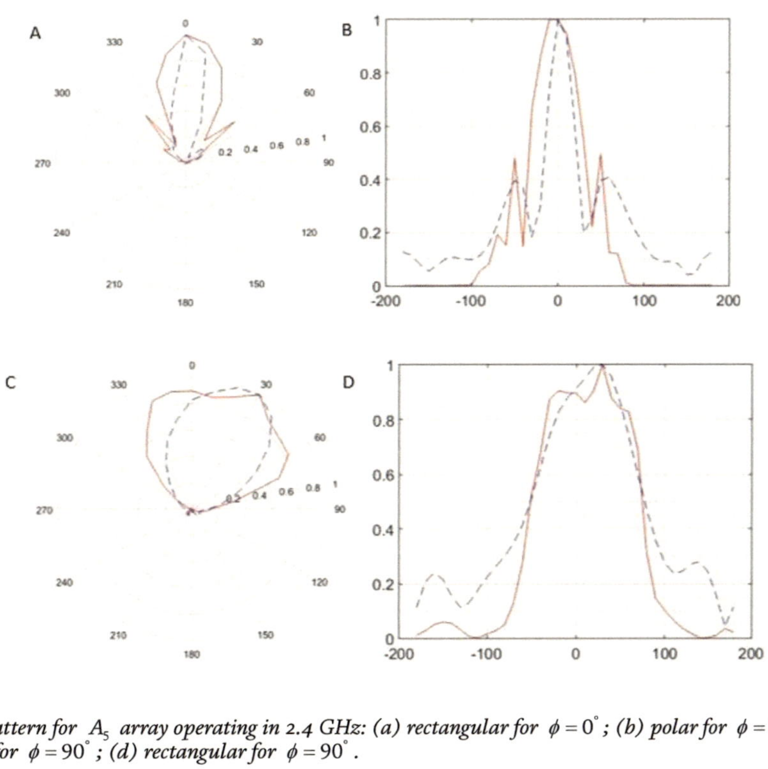

Figure 18.
Radiation pattern for A_5 array operating in 2.4 GHz: (a) rectangular for $\phi = 0^\circ$; (b) polar for $\phi = 0$; (c) rectangular for $\phi = 90^\circ$; (d) rectangular for $\phi = 90^\circ$.

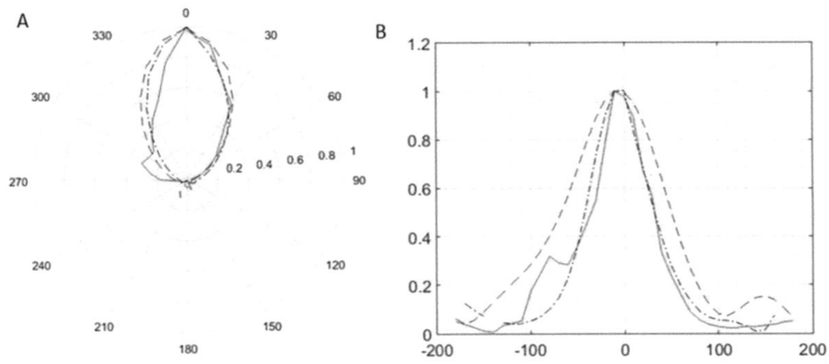

Figure 19.
Radiation pattern for A_4 *array operating in 2.4 GHz: (a) rectangular for* $\phi = 0^{\circ}$*; (b) polar for* $\phi = 0$.

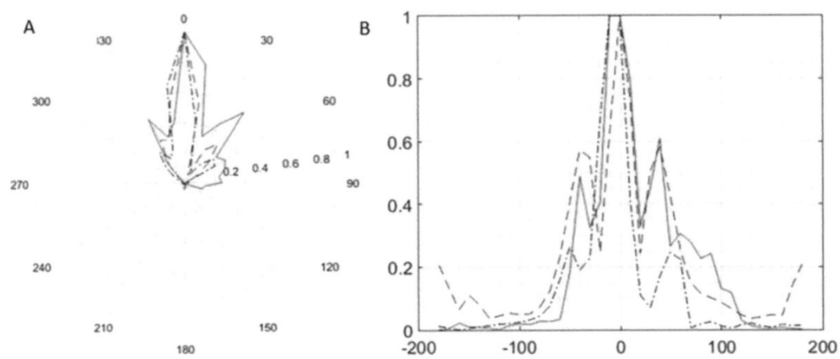

Figure 20.
Radiation pattern for A_6 *array operating in 2.4 GHz: (a) rectangular for* $\phi = 0$*; (b) polar for* $\phi = 0$.

Array	Measured Gain (dB)	Simulated Gain (dB)
A_1	6.145	7.027
A_2	6.17	6.94
A_3	7.06	7.21
A_4	3.165	1.43
A_5	7.86	8.00
A_6	9.76	10.20

Table 3.
Simulated and measured gain for each array.

From the radiation pattern measurements conducted for each of the arrays presented in this section, it is possible to calculate the gain of each model. The AUT gain is determined using the same measurement setup proposed and using (3).

$$P_{Rx} = P_{cal} + 2 * G_{AUT}(0^{\circ}) - L_{env} \tag{3}$$

Array	Measured $\phi = 0^\circ$	Simulated $\phi = 0^\circ$	Measured $\phi = 90^\circ$	Simulated $\phi = 90^\circ$
A_1	40°	70°	70°	70°
A_2	50°	60°	40°	70°
A_3	70°	70°	60°	90°
A_4	60°	80°	30°	40°
A_5	50°	40°	40°	40°
A_6	20°	20°	30°	40°

Table 4.
Measured and simulated HPBW results.

where P_{Rx} is the receiver power level measured in 0°, P_{cal} means the power measured using the calibration program, L_{env} is the free space loss determined by Friis [10, 25], and G_{AUT} is the measured gain. The simulated and measured gains are very similar, and only A_4 gain shows a higher difference due to the better return loss measured for this array, as can be seen in **Table 3**.

Half-power bandwidth (HPBW) is an important antenna parameter that shows the angle where the power decreases by about 3 dB. This parameter is related to AUT gain [8] and is obtained through radiation pattern analysis. HPBW results are shown in **Table 4**. The results prove the arrays are directive, as assumed for simulation, and therefore, they fit for use in nano cells in 5G services [26] and RFID applications.

4.2 Commercial wideband directional antenna

The information saved during the measurement included the antenna rotation angle and 50 samples of the received power. From this data, for each measurement angle, the average of the received power was calculated. Then, to verify the decay of

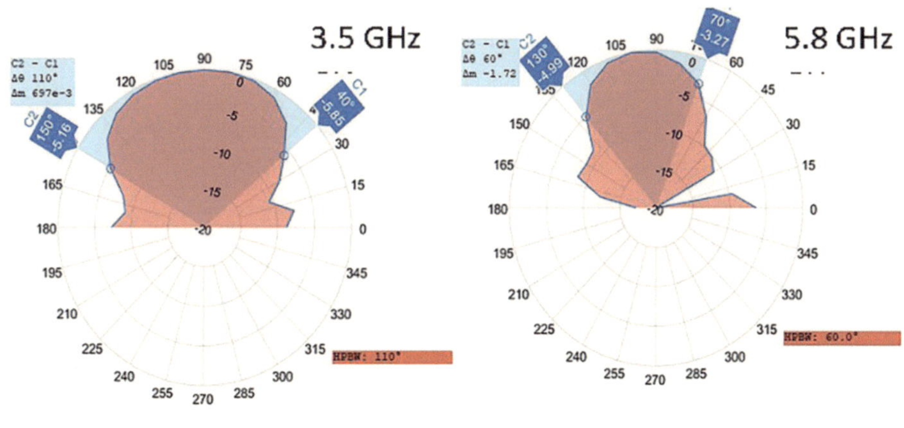

Figure 21.
Polar plot of the received power and HPBW calculation of the antenna for the frequencies of: (a) 3.5 GHz (b) 5.8 GHz.

the received power with angular variation, the power was normalized, and a polar plot was generated, as shown in **Figure 21**.

From the graph, it is possible to observe the decay of the received power with the angular variation of the antenna. To determine the HPBW of the antenna, the received power must vary by −3 dB relative to the direction of maximum radiation. In this case, for the frequency of 3.5 GHz, an HPBW of 110° is observed, and for the frequency of 5.8 GHz, an HPBW of 60°. To make the radiation pattern measurement more specific, the motor rotation can be reduced, for example, to 5°. The methodology described in the text was used for an commercial antenna purchased from online stores, for which radiation pattern information was not provided.

4.2.1 Measurements campaign

With the data generated through the use of the methodology for the antennas, a measurement campaign was conducted in an underground garage environment with the goal of determining the angular scattering of the received power in the 5.8 GHz range, for autonomous vehicle communication. The results of the measurement campaign were presented at the 16th Brazilian Congress of Electromagnetism through the paper [27].

In this work, a measurement campaign was carried out with fixed and mobile collections in an underground garage environment in the 5.8 GHz band. At each fixed measurement point, the angular scattering was analyzed to verify the influence of multipath components on the level of the received signal. Sixteen fixed points were measured, and the RX antenna was rotated at each 30° degrees. **Figure 22** presents a comparison between data collected with a directive antenna and an omnidirectional one. From the figure, it is possible to observe that for the first distance where the strongest component is the direct one and the components reflected by the nearby

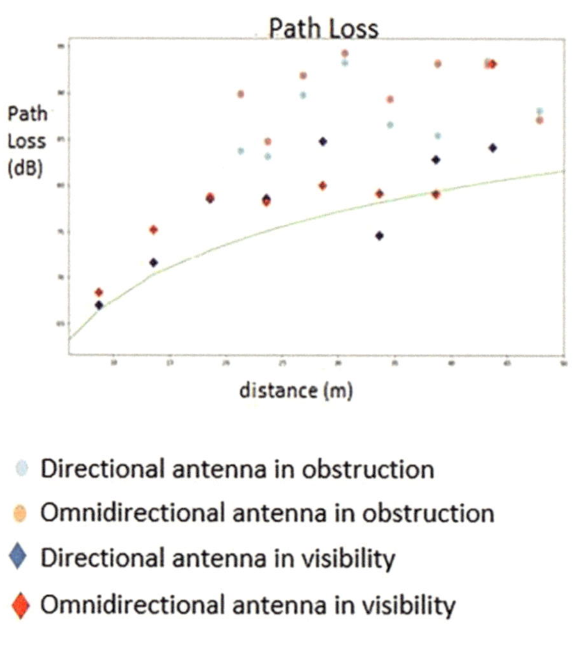

Figure 22.
Comparison of propagation loss calculated through overview of the components of the directional antenna, the antenna omnidirectional.

structures are scarce, the powers received from the omnidirectional antenna and the directional one are similar.

For distances further from the transmitter, the powers of the antennas are not more similar due to the difference in the path of each component that reaches the receiver, which, in the case of the omnidirectional antenna, can be constructive or destructive components. In the case of the directional antenna, each component is added to make up the total power received at each point.

5. Conclusions

A low-cost prototype was developed utilizing open-source software, Arduino, providing a practical and accessible solution for antenna characterization. The structural design of the prototype ensures portability due to its compact dimensions (40 x 5 cm), reduced height (less them 1 m), and lightweight construction using wood. The automated measurement system facilitates the real-time acquisition output signal parameters in equipment of the antennas radiation pattern characterization, with LabVIEW - MATLAB flow applications and analog digital converters. It enables the extraction of critical performance parameters for miniaturized high-frequency printed antennas at 2.4, 3.5, and 4.8 GHz carries, such as gain and half-power beamwidth (HPBW). The simulated results obtained are similar to the results measured in the indoor laboratory environment, proving the application of this technique. Measurements also carried out in an anechoic chamber validated the methodology used [28].

This system is particularly suitable for academic and research applications, addressing the challenges associated with the characterization of miniaturized antennas and can be applied to characterize even smaller antennas at higher frequencies above 5 GHz up to 60 GHz. Other different far-field antenna characterization techniques are addressed in the literature [29, 30] but they not considered the particularities of high-frequency miniaturized printed antennas as is proposed here, where the approach and prototype exhibit versatility, being applicable to a wide range of miniaturized antenna arrays and individual antenna elements operating at frequencies above 1 GHz without necessitating modifications to the system's structural configuration. Additionally, an alternative calibration methodology was introduced to validate the radiation pattern of a commercial directional antenna in 3.5 and 5.8 GHz by employing an open-area test site [27].

All the experimental and simulated results realized in this work confirm the feasibility of the proposed technique, which utilizes a pair of identical antennas without requiring the utilization a different standard antenna dedicated as reference at carrier frequency. Furthermore, the prototype and measurement technique have been successfully demonstrated in real-world scenarios, effectively mitigating the necessity for anechoic chambers or large open-field test environments, while ensuring accurate and efficient far-field antenna characterization radiation pattern and propagation loss in high-frequency applications.

Author details

Vanessa Przybylski Ribeiro Magri[1*], Leni Joaquim de Matos[1], Pedro Vladimir Gonzalez Castellanos[1], Vítor Luiz Gomes Mota[2] and Rodrigo Amitrano Bilobran[2]

1 Telecommunications Engineering Department (TET), Engineering School at Fluminense Federal University (UFF), Niterói, RJ, Brazil

2 Post Graduated University Program in Electrical and Telecommunications Engineering (PPGEET) at Fluminense Federal University (UFF), Niterói, RJ, Brazil

*Address all correspondence to: vanessamagri@id.uff.br

References

[1] Raaf B et al. Vision for beyond 4G broadband radio systems. In: 2011 IEEE 22nd International Symposium on Personal, Indoor and Mobile Radio Communications. Toronto, ON: IEEE Conference; 2011. pp. 2369-2373

[2] Andrews JG et al. What will 5G be? IEEE Journal on Selected Areas in Communications. 2014;**32**(6):1065-1082

[3] Qiao J, Shen XS, Mark JW, Shen Q, He Y, Lei L. Enabling device-to-device communications in millimeter-wave 5G cellular networks. IEEE Communications Magazine. 2015;**53**(1):209-215

[4] IEEE Spectrum. Everything You Need to Know About 5G. Available from: http://spectrum.ieee.org/video/telecom/wireless/everything-you-need-to-know-about-5g [Accessed: June 17, 2017]

[5] Andrews CJM, Narayanan ASK, Sunil AM. Compact metamaterial based antenna for 5G applications. Results in Engineering. 2024;**24**:103269. ISSN 2590-1230. DOI: 10.1016/j.rineng.2024.103269

[6] Jiang X, Deng S. Parameter research and design of 5G millimeter wave patch antenna. In: 2022 14th International Conference on Computer Research and Development (ICCRD). Shenzhen, China: IEEE Conference; 2022. pp. 405-410. DOI: 10.1109/ICCRD54409.2022.9730610

[7] Seddiki ML, Nedil M, Tebache S, Hadji SE. Compact multiband handset antenna design for covering 5G frequency bands. IEEE Access. 2024;**12**:20822-20829. DOI: 10.1109/ACCESS.2024.3362296

[8] Balanis CA. Modern Antenna Handbook. 1st ed. New York: John Wiley and Sons; 2005

[9] Krchnák A, Dzuriš M, Hartanský R, Halgoš J, Kliuchka A, Štibraný M. Investigation of the measurement distance for an UAV-performed radiation pattern measurement. Measurement Sensors Journal. 2024;**36**:101415. DOI: 10.1016/j.measen.2024.101415. ISSN26659174

[10] Balanis CA. Antenna Theory: Analysis and Design. New Jersey: John Wiley & Sons, Ltd; 2012

[11] Ortega A. Antenna performance evaluation through the lobular radiation pattern. In: Proc. of IEEE 4th Asia-Pacific Conference on Antennas and Propagation (APCAP). Bali Island, Indonesia: IEEE Conference; 2015

[12] Cajo R, Godoy C, Romero J, Bosquez C, Huilcapi V. Novel method to calculate antenna radiation patterns based on fixed-line-of-sight setting. In: Proc. of 27th IEEE International Conference on Microwaves, Antennas, Communication and Electronic Systems (COMCAS). Tel-Aviv, Israel: IEEE Conference; 2017

[13] Zwick T, Baks C, Pfeiffer UR, Liu D, Gaucher BP. Probe based MMW antenna measurement setup. In: Proc. of IEEE Antennas and Propagation Society Symposium. Monterey, USA: IEEE Conference; 2004

[14] Beer S, Zwick T. Probe based radiation pattern measurements for highly integrated millimeter-wave antennas. In: Proceedings of the Fourth European Conference on Antennas and Propagation. Barcelona, Spain: European Conference on Antennas and Propagation; 2010

[15] Taygur MM et al. Low-cost FPGA based antenna pattern measurement

system. In: Proc. of International Conference on Applied Electronics. Pilsen, Czech Republic: International Conference; 2013

[16] Hazan N, Sapuan SZ, Sayegh AM, Jenu MZM. Nasimuddin: "A portable measurement system for Antenna's radiation pattern". In: Proc. of IEEE Asia Pacific Microwave Conference (APMC 2017). Kuala Lumpur, Malaysia: IEEE Conference; 13-16 Nov 2017

[17] Brites FG, Santos VPA. Motor de Passo. Available from: https://www.telecom.uff.br/pet/petws/downloads/tutoriais/stepmotor/stepmotor2k81119.pdf

[18] Thomsem A. O que é Arduino?. Available from: https://www.filipeflop.com/blog/o-que-e-arduino

[19] Poole I. ELECTRONICS-NOTES (online). What is GPIB/IEEE 488 Bus. Available from: https://www.electronics-notes.com/articles/test-methods/gpib-ieee-488-bus/what-is-gpib-ieee488.php

[20] Wang J, Li G, Huang B, Miao Q, Fang J, Li P, et al. Spectrum Analysis and Regulations for 5G. Published in 5G Mobile Comunications Book. Springer; Oct 2016. pp. 27-50. DOI: 10.1007/978-3-319-34208-5_2

[21] Rodriguez V. Basic rules for anechoic chamber design, part one: RF absorber approximations. Microwave Journals. Jan 2016;**59**:4882

[22] Pozar DM, Schaubert DH. Microstrip Antenna: The Analysis and Design of Microstrip Antennas and Arrays. New York: IEEE Press; 1995

[23] Rea FM, Tovar N, Barba MF, Aviles JC. Approximate antenna radiation pattern determination using a low cost measuring system at 2.4 GHz. In: Proc. of IEEE 10th Latin-America Conference on Communications (LATINCOM), Guadalajara, Mexico. IEEE Conference; 14-16 Nov 2018

[24] Rheman A et al. Development of a cost effective antenna radiation pattern measurement setup. In: Proc. of 16th Mediterranean Microwave Symposium (MMS). Abu-Dabi, UAE: MMS Conference; 2016

[25] Garg R, Bhartia P, Bahl IJ, Ittipiboon A. Microstrip Antenna Design Handbook. Boston: Artech House; 2001

[26] Thompson J et al. 5G wireless communication systems: Prospects and challenges [Guest Editorial]. IEEE Communications Magazine. 2014;**52**(2):62-64

[27] Bilobran R et al. Characterization of angular scattering and signal propagation in the 5.8 GHz band for V2I communication in an underground garage environment. In: CBMAG 2024 - GOIANIA – GO, Brazil, 16th Brazilian Congress of Electromagnetism. CBMAG Conference; 2024 [Online]

[28] Mota VLG et al. Swivel low cost prototype and automatized measurement setup to determine 5G and RFID arrays radiation pattern. In: IEEE 3rd 5G World Forum, 5GWF. Bangalore, India: IEEE Conference Online; 2020

[29] Rammal Y et al. Measurement of the radiation pattern of a horn antenna in a vibrating intrinsic reverberation chamber. In: International Symposium on Electromagnetic Compatibility. Europe: EMC; 2023

[30] Van Den Biggelaar AJ et al. Method to experimentally verify far-field conditions in a direct far-field measurement setup. In: Antenna Measurement Techniques Association Symposium, AMTA. Daytona, Florida: Antenna Measurement Techniques Association Conference; 2021

Chapter 5

An Overview of THz Antenna Design for 5G/6G Wireless Communications

Haleh Jahanbakhsh Basherlou, Martins Odiamenhi, Naser Ojaroudi Parchin, Chan Hwang See and Ming Shen

Abstract

The rapid evolution of wireless communication technologies has spurred significant interest in terahertz (THz) frequencies as a key enabler for next-generation 5G and 6G wireless communications and mobile networks. THz antennas are critical for leveraging these high-frequency bands, offering unprecedented data rates, ultra-low latency, and enhanced spectral efficiency. This chapter provides an overview of various types of THz antennas designed for wireless applications, with a focus on their unique design requirements, fabrication techniques, and performance characteristics. Key topics include advances in antenna design, miniaturization, fabrication methods, and integration with electronic components. Additionally, the chapter explores the role of THz antennas in enabling essential array technologies, underscoring their transformative potential in next-generation wireless communication systems. Furthermore, the design details and fundamental characteristics of new THz antennas are discussed extensively, complementing and reinforcing the insights presented.

Keywords: 5G/6G, antenna design and miniaturization, antenna arrays, THz communications, wireless networks

1. Introduction

The evolution of wireless communication systems, encompassing both 5G and 6G, has continually pushed technological boundaries, striving for increased data rates, reduced latency, and enhanced connectivity. While 5G has already brought transformative improvements in speed, capacity, and reliability, 6G technology is anticipated to extend far beyond these advancements. Building on the progress of 5G, 6G aims to revolutionize a multitude of domains, including mobile communication, spacecraft communication, aircraft communication, submarine communication, and illumination communication [1, 2]. The envisioned 6G network seeks to provide coordinated and integrated coverage, driving global network performance enhancements. Both 5G and 6G address escalating industrial demands for high connection density, increased network efficiency, enhanced spectrum efficiency, and minimized end-to-end latency. Notably, in May 2018, the International Telecommunication Union (ITU) took a pivotal step by establishing an International Mobile Telecommunications (IMT) standard for 6G, targeting a potential launch by 2030 [3]. In parallel, the United States Federal Communication Commission (FCC) proposed the integration of 6G technology in THz spectrum-based networks and spatial multiplexing technologies during the Mobile World Congress in September 2018, highlighting global efforts toward next-generation communication systems [4].

The upcoming wireless communication systems signify a substantial leap in this technological evolution, targeting operational frequencies in the terahertz (THz) bands ranging from 0.1 to 10 THz, as shown in **Figure 1**. These frequencies correspond to wavelengths spanning from 3 mm to 0.03 mm, positioning 6G at the pinnacle of high-frequency communication technologies [5]. However, the transition to THz frequencies introduces a series of significant challenges, primarily due to the intrinsic characteristics of terahertz waves. The THz band, which occupies a radio frequency (RF) spectrum from 0.1 to 10 THz, offers two substantial advantages: abundant spectrum availability and the potential for broadband communication with extremely high data transmission rates. Yet, the high-frequency nature of the THz band also brings considerable limitations, such as high propagation losses, absorption, and scattering by atmospheric molecules and particles [6]. These challenges are particularly pronounced over extended distances, necessitating solutions to enhance communication range and reliability. One of the primary challenges associated with THz frequency bands, crucial for 6G development, is the high propagation loss, path loss, and atmospheric absorption caused by molecules. While 5G operates at lower frequencies that mitigate these effects, 6G aims to extend the boundaries of wireless communication by utilizing THz waves, which are significantly impacted by atmospheric conditions. Substantial absorption by water vapor and oxygen molecules, whose dimensions are comparable to THz wavelengths, leads to pronounced attenuation, severely limiting the effective communication range and coverage area [7, 8]. Overcoming these obstacles is essential for realizing the full potential of 6G wireless communication systems, building upon the foundation laid by 5G.

Wireless technologies operate across a broad spectrum of frequencies, each optimized for specific communication needs. Lower-frequency systems like GSM and NB-Internet of Things (IoT) provide long-range coverage suitable for mobile and IoT applications, but with lower data rates. Mid-range technologies such as LTE and WiMAX offer faster data rates over moderate distances, supporting broadband access and mobile communication. Short-range technologies like Wi-Fi, Bluetooth, and ZigBee enable high-speed communication in local environments, with Wi-Fi offering some of the highest data rates in this category [9]. As we move to higher frequencies, technologies like UWB, mmWave, and SATCOM offer ultra-high-speed data transfer,

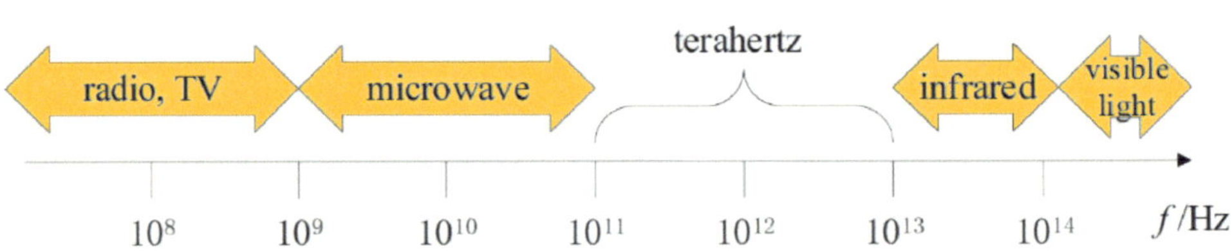

Figure 1.
The position of the THz wave in the electromagnetic spectrum.

with mmWave and Terahertz (THz) systems targeting even faster data rates for emerging applications. With the promise of speeds in the gigabit and terabit ranges, Terahertz communication, in particular, is poised to revolutionize wireless networks, requiring the development of advanced antenna designs capable of handling these ultra-high-frequency signals [10]. A detailed comparison of these technologies, including their frequency ranges, communication distances, and data rates, is provided in the **Table 1**.

The transition from 5G to 6G also introduces new complexities in antenna design and fabrication. While 5G technologies rely on advanced yet established antenna designs, the high frequencies and small wavelengths of the THz spectrum require entirely novel approaches. Conventional methodologies are inadequate for this spectrum, necessitating innovative solutions for efficient operation. Metamaterials, with their engineered electromagnetic properties, offer a promising avenue, enabling the creation of highly directional antennas with substantial gain to counteract the inherent limitations of THz communication. Addressing these challenges demands a multifaceted approach [11]. For both 5G and 6G systems, the development of antennas with high directionality or omnidirectionality and significant gain remains a critical focus. Research to optimize performance metrics—such as bandwidth, directivity, and gain—has resulted in advancements like the use of advanced substrate materials, corrosion-resistant designs, multi-element configurations, and dielectric lenses. Techniques to reduce mutual coupling, such as neutralization lines, decoupling networks, electromagnetic bandgap structures, defected ground structures, metamaterials, and slot elements, are increasingly vital for enhancing antenna performance in next-generation wireless systems [12].

This chapter provides an overview of the advancements in THz antenna design across both 5G and 6G wireless communication systems. It explores the unique issues posed by THz frequency bands for 6G and the latest innovations in design, fabrication, and measurement of antennas. This review covers both single-element and array

Technology	Frequency	Communication	Data rate
GSM	460 MHz–2 GHz	Up to 20 km	Up to 1 Mbps
NB-IoT	700–900 MHz	More than 10 km	Up to 200 Kbps
LoRaWAN	868 MHz–Sub-GHz	More than 10 km	Up to 50 Kbps
LTE Networks	1.9–2.1 GHz	Up to 5 km	Up to 100 Mbps
WiMAX Broadband	2.5–2.7 GHz	Up to 50 km	Up to 128 Mbps
Wireless Internet	2.4 GHz	Up to 30 m	Up to 150 Mbps
ZigBee IoT Protocol	868 MHz, 2.4 GHz	Up to 1 km	Up to 250 Kbps
Low-Energy Bluetooth	2.4 GHz	Up to 100 m	Up to 1 Mbps
WLAN	2.4 GHz, 5 GHz	Up to 100 meters	Up to 1.3 Gbps
UWB Systems	3.1–10.6 GHz	Up to 30 m	Up to 480 Mbps
mmWave Signals	24–100 GHz	Up to 100 meters	Up to 10 Gbps
SATCOM	1–40 GHz	More than 36 Mm	Up to 1 Gbps
Terahertz	100 GHz–10 THz	Up to 10 meters	10 to 160 Gbps

Table 1.
Comparison between various wireless technologies.

antennas reported for THz applications. In addition, two specific antenna designs including single-element and an array THz antenna are presented, along with a detailed discussion of their characteristics.

2. Development of THz antennas

The design and development of THz antennas are pivotal in advancing modern communication technologies, bridging the capabilities of 5G, and paving the way for 6G networks. THz antenna design is at the forefront of enabling cutting-edge applications such as ultra-HD streaming, immersive virtual and augmented reality, autonomous vehicle networks, advanced sensing, and IoT-based smart systems. This section provides an exploration of various THz antenna designs, their unique characteristics, and their transformative role in shaping the future of communication technologies. Each type of antenna, whether planar or 3D, single-element or array, carries unique design considerations tailored to specific applications, enabling diverse functionalities and efficiencies for present and future wireless networks [13].

2.1 Single-port/single-element THz antennas

The Single-port/single-element THz antennas represent a crucial frontier in the development of 6G networks, primarily due to their capability to harness the vast bandwidths available in the terahertz frequency spectrum (0.1 to 10 THz). Their compact size, enabled by the short wavelengths at THz frequencies, facilitates integration into small form factor devices, paving the way for ubiquitous connectivity on the Internet of Things (IoT) era. However, their deployment faces significant technical challenges, including precise fabrication to maintain performance at THz frequencies, mitigating losses associated with material characteristics, and overcoming integration hurdles with existing electronic systems [14]. Recent advancements leverage novel materials such as graphene and metamaterials, which offer unique properties like tunability and reduced signal loss, thereby enhancing antenna efficiency and performance. This sub-section discusses the various designs of single port and single element THz antennas along with their outcomes.

2.1.1 Microstrip/printed THz antennas

Microstrip-printed antennas, renowned for their compact form factor, ease of fabrication, and seamless integration with planar circuits, have become pivotal in leveraging this spectrum. One notable advancement in the realm of THz antennas is the development of a wideband substrate integrated waveguide (SIW) based slot antenna designed for D-band applications (0.11–0.17 THz) [15]. This antenna employs a two-step wideband WR6-SIW transition to ensure effective impedance matching and accurate measurement. The design achieves an impressive −10 dB impedance bandwidth of 42.86%, spanning from 0.11 to 0.17 THz, a significant bandwidth enhancement achieved by merging six closely spaced resonance frequencies. This intricate mode merging strategy not only broadens the bandwidth but also enhances the antenna's radiation characteristics, making it a viable candidate for future 6G applications. Another significant contribution is the use of microstrip patch antenna (MPA) operating at 0.835, 0.635, and 0.1 THz frequencies, designed on a liquid crystalline polymer substrate using a simple PCB etching process [16]. These antennas are targeted for THz spectroscopy in cancer detection and Doppler radar-based vital sign detection, demonstrating the versatility of THz applications beyond communication. The fabricated prototype array at 0.1 THz exhibited good agreement between measured and simulated results, showcasing the approach's potential for practical implementation. The study highlights the improved gain and fabrication tolerance achieved with the liquid crystalline polymer substrate. The development of a quasi-Yagi antenna designed for D-band frequencies (0.11–0.17 THz) integrated within a glass interposer represents another innovative approach to THz antenna design [17]. This antenna offers a wide impedance bandwidth and a peak gain of 4.78 dBi at 0.14 THz, making it highly suitable for 6G handset applications. The use of a glass interposer facilitates compact integration, leveraging advanced glass panel embedding techniques. One of the key challenges addressed is the measurement of end-fire antennas above 100 GHz, with a proposed probe-station-based setup for return loss and end-fire gain measurements. Additionally, an envelope detector circuit is utilized for characterizing the normalized radiation pattern, demonstrating consistency with simulation results. The integration of the quasi-Yagi antenna in a glass interposer provides a promising antenna-in-package solution, capitalizing on the beneficial properties of glass substrates, such as low loss and high integration density, critical for 6G applications., the photonic crystal (PC) structures for designing a microstrip patch antenna operating in the terahertz (THz) frequency range (0.1–1 THz) is investigated in [18]. An inset-feed rectangular microstrip antenna utilizing 1D, 2D, and 3D PC substrates has been proposed. The study evaluates various PC materials like air cavity, polyamide, paper, FR4, Arlon, and quartz, highlighting the superior efficiency of air cavity PBG structures. Antennas based on 2D PCs demonstrate impressive performance, achieving a minimum reflection coefficient of −63.22 dB, maximum directivity of 6.81 dBi, and radiation efficiency of 88.17% at 0.741 THz. Simulation using CST Microwave Studio includes analysis of scattering parameters and material properties, concluding that 2D PC antennas with air cavities are optimal for the 0.68–0.74 THz frequency band, suitable for applications in threat detection, homeland security, and wireless communication.

2.1.2 3D/non-planar THz antennas

The advancements in 3D and non-planar THz antennas highlight significant strides in achieving the performance metrics required for 6G networks. These innovations are poised to facilitate the integration of THz technologies into future 6G communication systems, supporting a wide array of applications from high-speed data transmission to advanced sensing. One notable advancement in this field is the development of 0.3-THz step-profiled corrugated horn antennas designed for integration into low-temperature co-fired ceramic (LTCC) packages [19]. These antennas utilize substrate-integrated waveguide technology to form a hollow waveguide and horn structure within a multilayer LTCC substrate, surrounded by a *via* fence, for enhanced performance. Experimental results show the LTCC waveguide demonstrating low insertion loss of 0.6 dB/mm, while the LTCC horn antenna achieves an 18-dBi peak gain with a 0.1-THz bandwidth and over 10-dB return loss. The compact design facilitates seamless integration into LTCC transceiver modules, with experimental results aligning closely with simulations. Another significant contribution is the introduction of a dielectric rod waveguide (DRW) antenna designed for frequencies ranging from 0.075 to 0.325 THz [20]. The antenna's broadband geometry is optimized through numerical simulations and is matched with metal waveguides of varying sizes across different frequency bands. Measurements confirm close agreement with simulations up to 0.325 THz, maintaining nearly constant gain across the entire frequency range with a relative bandwidth of 160%. However, sharp antenna tips limit performance beyond 0.325 THz due to manufacturing constraints. The antenna demonstrates better than 15 dB return loss, with radiation patterns nearly independent of frequency. The development of a suspended SOG tapered antenna operating in the frequency band of 0.11–0.13 THz represents another innovative approach in THz antenna design [21]. This antenna, fabricated using deep reactive ion etching (DRIE) of the Si layer and selective etching of the glass substrate, radiates both Ex- and Ey-polarizations. Key features include linear polarization, high directivity, integrability, and low-cost fabrication. Experimental results validate its performance, showing gains of 17.5 dBi at 0.115 THz for Ex-polarization and 18.3 dBi at 0.128 THz for Ey-polarization, with measured efficiencies of 94% and 99%, respectively. The suspended SOG tapered antenna enhances radiation efficiency and gain in mmW applications, designed, fabricated, and measured using precise photolithography and DRIE processes. This configuration achieves high directivity and efficiency, making it suitable for transceiver applications in the 0.11–0.13 THz range. In [22], the use of metallic 3D printing for antennas operating up to 0.325 THz is discussed which compares binder jetting/sintering on stainless steel and selective laser melting on Cu–15Sn, ultimately selecting Cu–15Sn for its cost-performance balance. This led to the development of conical horn antennas for the E-, D-, and H-bands, which demonstrated strong agreement between simulated and measured performance, covering the entire operational band with gains over 21.5 dBi. The study highlights that compared to traditional methods, 3D printed antennas offer environmental benefits, lower costs, and faster production times. Additionally, these metallic 3D printed antennas provide greater simplicity and robustness than their non-metallic counterparts, underscoring the potential of metallic 3D printing for both industrial mass production and prototyping.

2.1.3 On-chip THz antennas

The development of on-chip antennas is critical for the miniaturization and integration of THz technologies in 6G wireless systems. On-chip antennas offer the potential for high performance and compact size, essential for advanced communication and sensing applications. The advancements in on-chip THz antennas highlight significant strides in achieving the performance metrics required for 6G networks. These innovations are poised to facilitate the integration of THz technologies into future 6G communication systems, supporting a wide array of applications from high-speed data transmission to advanced sensing [23]. This section explores recent progress in on-chip THz antennas, focusing on their innovative designs, fabrication techniques, and potential impact on future 6G networks. A significant advancement in on-chip THz antennas is the development of antennas operating at 0.165 THz, designed using a standard silicon-Germanium BiCMOS process with localized backside etching (LBE) to create air trenches in silicon [24]. This study developed and characterized three antennas in the D-band (0.11–0.17 THz): a dipole antenna optimized for a tilted beam achieving 1 dBi gain, an LBE-based folded dipole achieving 5 dBi gain at 0.165 THz over 1.88 mm^2, and an LBE-patch antenna attaining 6 dBi gain at 0.16 THz over 1 mm^2. The study emphasizes optimizing antenna geometry for process reliability and examines the effects of metal fillings on radiation patterns and matching. Another notable development is the introduction of a 0.45-THz on-chip antenna designed for wide bandwidth and an exceptionally low profile [25]. Utilizing a dual-patch structure, this antenna achieves an impedance bandwidth exceeding 15% at a profile height of just 0.013 $\lambda 0$. Simulation results indicate a peak gain of 2.7 dBi and a radiation efficiency of 31.7%. Fabricated using 65-nm CMOS technology, preliminary measurements confirm an impressive 15.9% impedance bandwidth, consistent with simulations. The antenna's characteristics, including its low profile and wide bandwidth, make it suitable for future applications in 6G wireless systems, short-range communications, and terahertz detection. Enhancing the performance of a 0.3-THz on-chip patch antenna by placing it in a low-cost quad-flat no-lead (QFN) package using silica-based materials represents another significant advancement [26]. Full-wave simulations for a rectangular patch antenna, designed to match a 65 nm CMOS process, reveal that when the packaging material thickness over the antenna is approximately $\lambda/3$, radiation efficiency improves from 46–60% at 0.3 THz, with a 7 GHz bandwidth increase and 1 dB peak gain. Practical tests with a 0.276-THz CMOS signal generator equipped with the on-chip antenna in a QFN package show an effective isotropic radiated power (EIRP) about 6 dB higher than unpackaged versions. This improvement is attributed to enhanced antenna performance due to the packaging. A THz CMOS on-chip patch antenna with a defected ground structure (DGS), designed to achieve broadband and high gain is proposed in Ref. [27]. Simulation results show that the DGS enhances the antenna element's gain, bandwidth, and isolation. The antenna is fabricated using a commercial 65 nm CMOS process and measured on-wafer. They exhibit gains of 3.1 dBi, with bandwidths of 14.0%, for a reflection coefficient less than −10 dB at 0.3 THz. The CMOS on-chip antenna designed in this work promises low-cost and high-performance integration into THz systems using standard CMOS processes without additional manufacturing techniques.

2.1.4 Summary and comparison

Table 2 provides a summary and comparative overview of single-port/single-element THz antenna designs, focusing on their fabrication methods, operational frequency ranges, and unique characteristics. Each design showcases innovations aimed at overcoming the technical challenges of THz antenna deployment, such as precise fabrication, efficient material utilization, and integration compatibility. For instance, microstrip antennas demonstrate advancements in substrate technologies, like liquid crystalline polymers and photonic crystals, to enhance bandwidth and efficiency. Similarly, 3D and non-planar antennas leverage cutting-edge methods such as LTCC and metallic 3D printing to achieve high gain and compactness, essential for integration into transceiver modules. On-chip antennas highlight the potential for miniaturization, employing advanced CMOS and BiCMOS processes to enable compact and high-performance designs suitable for 6G systems. The diversity of these designs underscores the ongoing efforts to address specific requirements, including wide bandwidth, high efficiency, and cost-effective manufacturing, paving the way for practical THz communication and sensing applications.

2.1.5 Simulations and design example of a single-element THz antenna

In this sub-section simulation and design details and radiation characteristics of a single-element THz antenna are discussed. The properties of the design were examined using CST 2022 [28]. **Table 3** provides the specific parameter values for

References	Antenna type	Fabrication method	Frequency (THz)	Main characteristics
[15]	SIW Slot Antenna	Substrate Integrated Waveguide (SIW)	0.11–0.17	Wide bandwidth; merges six modes.
[16]	Liquid Crystal Polymer MPA	PCB etching on liquid crystalline polymer	0.1, 0.635, 0.835	High fabrication tolerance; versatile for spectroscopy and radar.
[17]	Quasi-Yagi Antenna	Glass interposer with panel embedding	0.11–0.17	Compact integration; probe-based end-fire.
[18]	Photonic Crystal MPA	Inset-feed with PC substrates	0.68–0.74	Air cavity PC substrates; high efficiency.
[19]	Corrugated Horn Antenna	LTCC with hollow waveguide	0.1–0.3	High gain; compact integration modules.
[20]	Dielectric Rod Waveguide	Metal waveguide with optimized geometry	0.075–0.325	Broadband with stable gain across frequencies.
[21]	Suspended SOG Tapered	Photolithography and DRIE	0.11–0.13	Dual polarization; high efficiency.
[22]	3D Printed Conical Horn	Metallic 3D printing (Cu–15Sn)	Up to 0.325	Cost-effective and eco-friendly fabrication.
[24]	On-Chip Dipole	LBE in SiGe BiCMOS	0.11–0.17	Compact size; optimized for process reliability.
[25]	On-Chip Dual-Patch Antenna	CMOS 65-nm	0.45	Low profile; wide impedance bandwidth.
[26]	Packaged Patch Antenna	CMOS in QFN package	0.276–0.3	Enhanced efficiency and gain with packaging.
[27]	On-Chip Patch	CMOS 65-nm with DGS	0.3	Improved isolation, bandwidth, and gain.

Table 2.
Summary of the discussed single-element THz antennas.

Parameter	W_X	L_X	W_1	L_1	W_2	L_2
Value (μm)	70	60	12	5	5	10
Parameter	L_3	W_4	L_5	W_5	W_6	W_3
Value (μm)	5	25	15	5	25	7.5

Table 3.
Parameter values of the designed THz antenna.

the suggested antenna design. **Figure 2(a)** illustrates the configuration and design details of the single element S-shaped monopole resonator which characterized by its compact dimensions, printed in a 5 μm thick Rogers RO3003 material. The choice of the S-shaped monopole design is driven by its compact size, dual-band functionality, make it suitable for modern devices of the next-generation wireless communication systems. **Figure 2(b)** presents the S_{11} (reflection coefficient) result showing that the antenna impedance bandwidth spanning from 2.45 to 2.55 THz and 3.3 to 3.75 THz at a − 10 dB threshold within the 6G spectra.

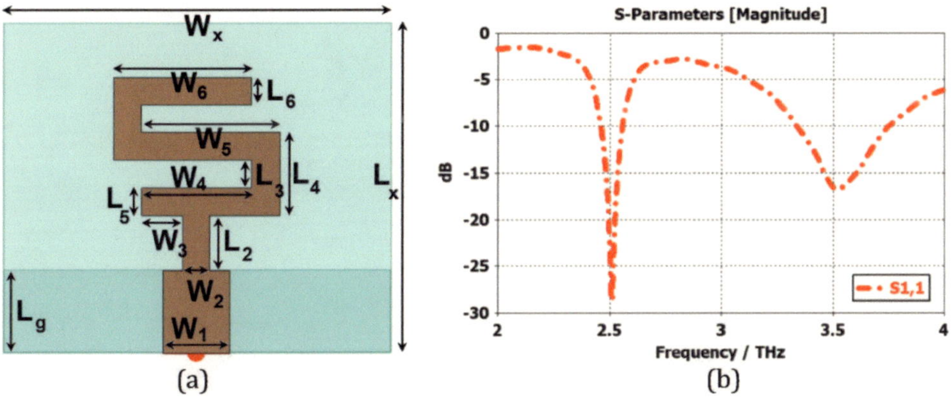

Figure 2.
(a) Single-element design details and (b) its S_{11} result.

To theoretically discuss and validate the dual-band functionality of the proposed antenna, the simulated current distributions at the resonance frequencies of 2.5 THz and 3.5 THz are presented in **Figure 3**. Notably, during the first resonance at 2.5 THz, the longer section of the S-shaped structure exhibits significant current flow. Conversely, in the second resonance at the higher frequency of 3.5 THz, it is primarily the lower (or half) section of the S-shaped antenna that demonstrates considerable current densities. This behavior underscores the inverse relationship between antenna size and frequency, where shorter dimensions correspond to higher frequencies [29].

Figure 4 illustrates the efficiency results across the resonance frequencies. As depicted, the element exhibits high-efficiency rates throughout its operational bandwidth. Radiation efficiencies exceed 95%, while total efficiencies remain above 85% across the two frequency bands. Notably, within the mid-frequency range, both values surpass 95%, demonstrating that the proposed design is suitable and reliable for 6G communications. Furthermore, the gain results presented in **Figure 5** indicate that the suggested monopole design achieves significant gain levels, enhancing its performance.

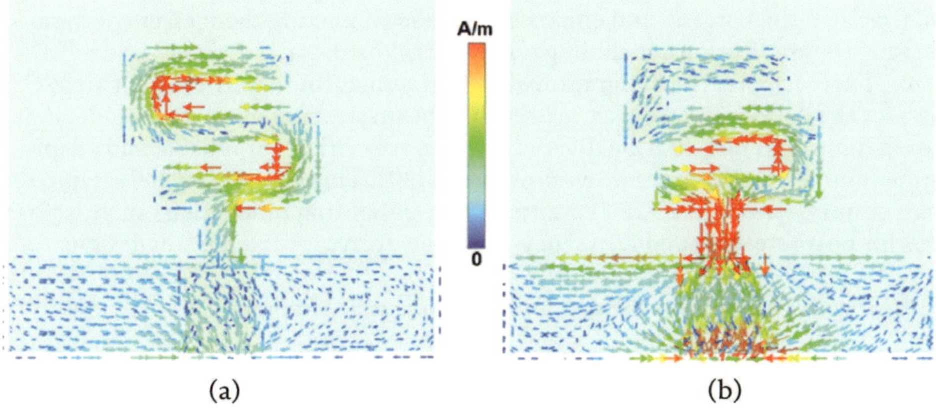

Figure 3.
Current distributions at (a) 2.5 and (b) 3.5 THz.

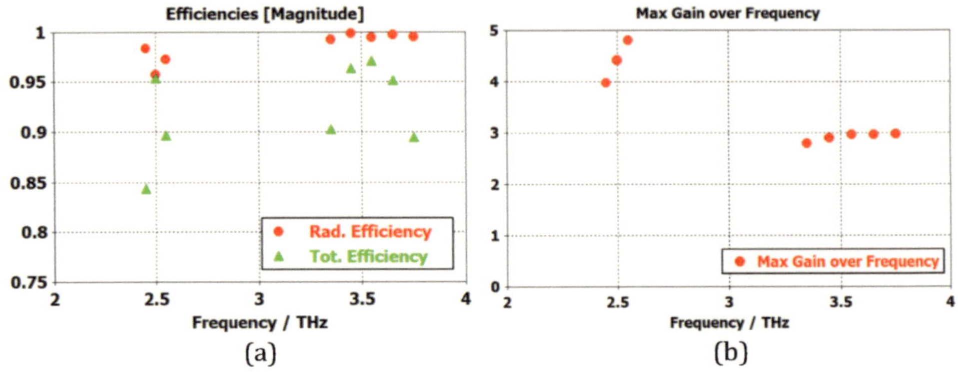

Figure 4.
(a) Efficiencies and maximum gain results of the designed THz antenna.

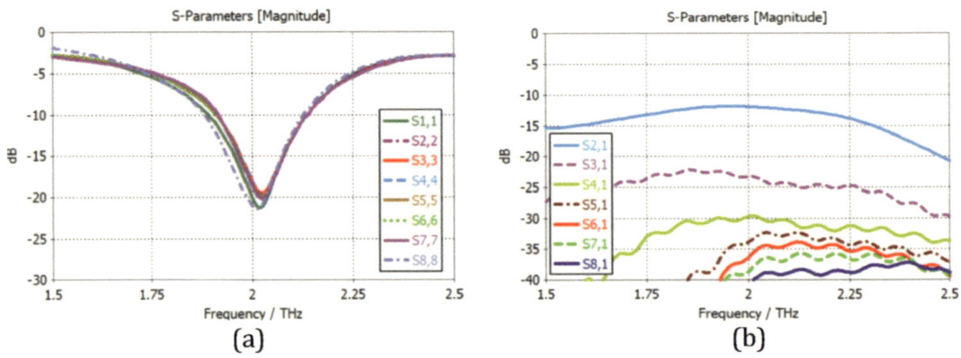

Figure 5.
Scattering parameters: (a) S_{nn} and (b) S_{n1}.

2.2 Multi-prot/array THz antennas

Multi-port and array THz antennas are pivotal in addressing the growing demands of high-capacity, high-speed communication networks, especially for 6G systems. These antennas leverage array configurations and multi-port designs to achieve high gain, precise beam control, and enhanced efficiency, making them essential for a wide range of applications, including wireless backhaul, radar systems, and advanced sensing. Furthermore, the integration of THz antennas into multi-element arrays allows for the development of features such as beam steering, frequency-dependent beam shaping, and polarization diversity, which are critical for reliable and adaptable communication in dynamic environments [30]. This sub-section delves into the design of multi-port and array THz antennas, highlighting innovations such as phased arrays for beam steering and frequency-scanning arrays for fixed beam designs.

2.2.1 High-gain THz antenna arrays: Beam-steerable phased arrays

Phased array antennas with beam-steerable capabilities at THz frequencies are critical for achieving high data rates, precise radiation steering, and enhanced connectivity required for 6G wireless networks. A notable development in phased array THz antennas is the 0.37–0.41 THz phased-array transmitter utilizing W-band components and an eight-element quadrupler array connected to high-efficiency microstrip antennas built with CMOS technology [31]. The design's scalability, leveraging W-band frequencies, avoids high transmission-line loss at 0.4 THz. This represents one of the first demonstrations of a CMOS-based phased array operating at such high frequencies with a wide bandwidth, highlighting its potential for scalable and efficient THz communications in 6G networks. Another contribution is the 0.14-THz wideband array antenna-in-package (AiP) designed for compatibility with flip-chip technology and integrated transceivers [32]. Utilizing multimode resonance on a low-profile multilayer PCB, this design incorporates multiple resonances from a patch and $\lambda/4$ monopole-type feeder, achieving a simulated impedance bandwidth of 53% with stable radiation performance. A 4 × 4 antenna array demonstrates up to 18.1 dBi gain, 80% radiation efficiency, and over 20 dB cross-polarization discrimination (XPD). Experimental results show a measured −10 dB impedance bandwidth of 31%, making this AiP solution a promising candidate for high-performance 6G applications, balancing wide bandwidth and compact form factor. The development of a phased array solution for ultra-sharp beam forming and high-angular-resolution steering at 0.265 THz represents another breakthrough [33]. This approach reduces the required aperture size for a 1° beamwidth, facilitating implementation with CMOS microelectronic chips. A 98 × 98 antenna element array demonstrates the formation and electronic steering of a THz pencil beam with approximately 1° beamwidth in two dimensions. Using a 1-bit phase-shifting reflective antenna with cross-polarization backscattering, the design achieves precise 0°/180° phase inversion and maintains performance despite quantization errors. This method effectively reduces sidelobe and squint, supporting monolithic integration for advanced THz applications. The high-angular-resolution steering and ultra-sharp beam forming capabilities make this design highly suitable for precise and high-capacity 6G wireless systems. Additionally, a 0.28-THz phased array transmitter featuring an integrated silicon-based antenna offers another innovative solution for THz applications [34]. Developed using a 65-nm CMOS process, this transmitter achieves a peak EIRP of 9.3 dBm with a 3-dB bandwidth of 20 GHz. The design integrates a 4 × 4 phased array, employing phase shifters and power amplifiers to enhance beam steering and signal amplification. The antenna array demonstrates a beam-steering range of ±30°, with measured results closely aligning with simulations. A promising approach for THz photonic circuits involves micro-scale silicon photonic crystal waveguides enhanced by monolithically integrated gradient-index (GRIN) optics [35]. Integrating gradient-index (GRIN) optics with silicon photonic crystal waveguides, recent research presents two innovative devices: a Luneburg lens-based multi-beam antenna and a Maxwell fisheye lens-based slab-mode beam launcher [36]. These lenses show great potential for THz antennas, multiplexers, and power-combining devices. Furthermore, a seven-port multi-beam antenna, implemented with a GRIN Luneburg lens coupled to an array of photonic crystal waveguides, demonstrates significant potential for dense communication networks and directionally aware short-range radar.

2.2.2 High-gain THz antenna arrays: Fixed and frequency-dependent beam angles

THz antenna arrays designed for fixed beams or with frequency-dependent beam angles are crucial for high-gain applications in 6G wireless networks, offering robust performance without the complexity of active phase shifting or beam steering mechanisms. This sub-section reviews recent advancements in such antenna arrays, emphasizing their fabrication techniques, gain characteristics, and potential for 6G applications. A significant advancement in high-gain THz antennas is the development of high-gain antennas with broad bandwidth for the 0.12 THz band, fabricated by diffusion bonding of laminated thin copper plates [37]. This approach offers high precision and low loss at high frequencies. The design includes a double-layer feeding structure ensuring stable fabrication. A 32 × 32-element array antenna achieves a 38 dBi gain with 60% efficiency over a 0.015 THz bandwidth (0.119–0.134 THz), while a 64 × 64-element array achieves a 43 dBi gain with 50% efficiency over a 0.0145 THz bandwidth (0.1185–0.133 THz). These low-profile antennas could serve as alternatives to conventional high-gain antennas like reflector and lens antennas, making them suitable for 6G applications that require high efficiency and compact design. Innovative design techniques for terahertz antenna-in-package (AiP) systems have been proposed to address challenges in integration, fabrication, and measurement of multilayer PCB-based antenna arrays [38]. Key innovations include a wideband dual-polarized stub-loaded proximity-coupled stacked patch antenna and a compact vertical power divider. These solutions were validated through circuit analysis and demonstrated with a 4 × 1 subarray, paving the way for an 8 × 8 AiP for future 6G communications. The stub-loaded antenna enhances bandwidth using open and short stubs *via* transitions, while the vertical power divider simplifies the feeding network. Measurement of the 8 × 8 antenna array package showed a boresight gain exceeding 17.1 dBi across the 0.136–0.148 THz bandwidth, highlighting its potential for robust and efficient 6G communication systems. Another significant contribution is the development of frequency scanning slot arrays operating from 0.13 to 0.18 THz, micro-fabricated using the PolyStrata sequential copper deposition process [39]. The voltage standing wave ratio is less than 1.75:1 over the entire range, with measured scanning of 0.00104 THz from 0.13 to 0.15 THz and 32.5° over the full range. A 10-element array achieves a gain of 15.5 dBi at 0.15 THz, and a 20-element array achieves 18.9 dBi at 0.15 THz, with about 3 dB variation over the scan range. The results align with HFSS full-wave simulations, demonstrating effective beam scanning and gain performance. These slot arrays are well-suited for applications requiring high-gain and frequency-dependent beam angles in the THz range. A novel architecture for a sub-THz antenna-in-package (AiP) enhances isolation between ports in a dual-polarized stacked patch antenna [40]. It integrates orthogonal fan-out lines from a vertical *via* for probe feeding, augmented with grounded shielding structures. This design achieves up to 10 dB isolation improvement per polarization. Validation through simulations and measurements of a 4 × 1 subarray shows an S21 value of −20 dB at 0.145 THz, with a measured S11 bandwidth exceeding 0.01 THz and return loss of 10 dB. The proposed structure enhances AiP performance for sub-THz applications, ensuring stable communication systems with improved port isolation, essential for high-performance 6G networks. The design and development of a novel frequency beam-scanning array antenna operating in the Y-band (0.22–0.325 THz) represent another innovative approach [41]. The antenna features a traveling-wave structure with a meandered rectangular waveguide and a slot-coupled cavity-backed patch array for elevation. It achieves a narrow beamwidth of ~2.5° and ± 25° beam steering from 0.23 to 0.245 THz, with a 10° elevation beamwidth. The array, consisting of over 600 patch elements, provides a gain of over 29 dBi and radiation efficiency above 55%, all within a compact 45 mm × 8.5 mm × 1.25 mm, 4.5 g structure. Fabricated using silicon micromachining, the prototype demonstrates a measured scanning range of over 48° and gain over 28.5 dBi, aligning well with simulations. This design is ideal for low-mass, compact Y-band radar applications, offering high gain and precise beam control. A corporate-feed slotted waveguide array antenna for the 0.35-THz band has been designed and fabricated using the DRIE process to achieve high fabrication accuracy [42]. The thin laminated plates forming the antenna were etched with tolerances lower than ±5 μm and bonded using diffusion bonding. The gold-plated silicon wafer showed an effective conductivity of 1.6×10^7 S/m and a loss per unit length of 1.1 dB/cm. The 16 × 16 element array antenna demonstrated a gain of 29.5 dBi at 0.35 THz, with a 3-dB bandwidth of 0.0508 THz in simulations and 0.0446 THz in measurements. This is the first demonstration of a broadband antenna in this frequency band, suitable for applications such as short-range broadband wireless communication, highlighting its potential for high-frequency, high-gain 6G networks.

2.2.3 Summary and comparison

Table 4 provides a comprehensive overview of advancements in multi-port and array THz antennas, highlighting their diverse fabrication methods, frequency ranges, and unique characteristics tailored for 6G applications. CMOS-based phased-array transmitters offer scalability and low loss with wide bandwidths, while AiP solutions utilizing multimode resonance provide high gain and wide impedance bandwidth. Ultra-sharp beamforming antennas and integrated silicon-based phased arrays enhance angular resolution and beam control. GRIN optics-based antennas are suited for multi-beam applications, while laminated copper plate antennas excel in compact, high-precision designs. Frequency-scan slot arrays and stub-loaded AiP antennas cater to frequency-dependent beam steering, and novel sub-THz AiP and meandered waveguide antennas offer compact, high-gain solutions. Corporate-feed slotted waveguide arrays provide high accuracy and broadband performance. These designs reflect the diverse approaches to meeting the advanced requirements of emerging 6G networks.

2.2.4 Simulation and design details of phased array THz antenna

This sub-section explores the characteristics of a linear phased array designed to provide high gain and steerable radiation patterns, tailored to meet the demands of upcoming 6G networks. **Figure 6** displays the configuration of an 8-element array, arranged in a 1 × 8 layout with an element spacing of d = 80 μm. The array features a schematic with eight modified dipole elements arranged in a 1 × 8 linear format with overall dimension of $W_a \times L_a = 480 \times 70\ \mu m^2$. To improve the radiation efficiency of the dipole resonators, rectangular directors are strategically placed adjacent to each resonator. These compact resonators utilize microstrip-line feeds, optimizing their design.

The S-parameter results (S_{nn}/S_{n1}) for this array are shown in **Figure 5**, highlighting its operational coverage from 1.9 to 2.1 THz, encompassing critical frequencies within the emerging 6G and beyond spectrum. The results illustrate that the resonators exhibit low mutual coupling, consistently below −12 dB across the entire bandwidth. When evaluating the gain levels, as shown in **Figure 7(a)**, the single antenna achieves a gain ranging from 4.5 to 5.5 dBi. In contrast, the linear array shows significantly higher gain levels, varying between 12.5 and 14.5 dBi. Notably, the array exhibits a clear trend of increasing gain as the operating frequency rises, which highlights the array's capability to enhance performance at higher frequencies. This gain improvement, coupled with the consistent end-fire radiation, underscores the array's suitability for high-frequency THz applications, offering robust signal strength and wide-area coverage necessary for next-generation systems.

References	Type of antenna	Fabrication method	Frequency (THz)	Main characteristic
[31]	CMOS-based phased-array transmitter	CMOS technology	0.370–0.410	Scalability, wide bandwidth, low transmission-line loss
[32]	AiP with multimode	Flip-chip, multilayer PCB	0.140	High gain, wide bandwidth (53%)
[33]	Ultra-sharp beamforming	CMOS microelectronic chips	0.265	High-angular-resolution, sharp beam forming, reduced sidelobe/squint
[34]	Integrated silicon-based antenna	65-nm CMOS process	0.280	Peak EIRP, ±30° beam-steering, amplification
[35]	GRIN optics-based	integration of gradient-index (GRIN) optics	Various (THz range)	Multi-beam formation, potential for dense and radar applications
[36]	Laminated copper plates	Diffusion bonding of laminated thin copper plates	0.119–0.134	High precision, low-loss, high gain up to 43 dBi, compact design
[37]	Stub-loaded AiP antenna	multilayer PCB, compact divider	0.136–0.148	Improved port isolation, wide bandwidth, high gain
[38]	Frequency-scan slot array	PolyStrata sequential copper deposition	0.130–0.180	Effective scanning, high gain, compact and efficient design
[39]	Sub-THz AiP with enhanced isolation	Stacked antenna orthogonal fan-out and shielding	0.145	Improved isolation, stable performance, robust for sub-THz
[40]	Meandered rectangular waveguide	slot-coupled cavity-backed array	0.230–0.245	High gain, low mass, compact for radar applications
[41]	Corporate-feed slotted waveguide	DRIE process, laminated thin copper plates	0.350	High fabrication accuracy, broadband performance, high gain
[42]	Frequency-scan slot array	PolyStrata sequential copper deposition	0.130–0.180	Efficient beam scanning, High gain, suitable for frequency-dependent beam angle applications

Table 4.
Summary of the discussed THz array antennas.

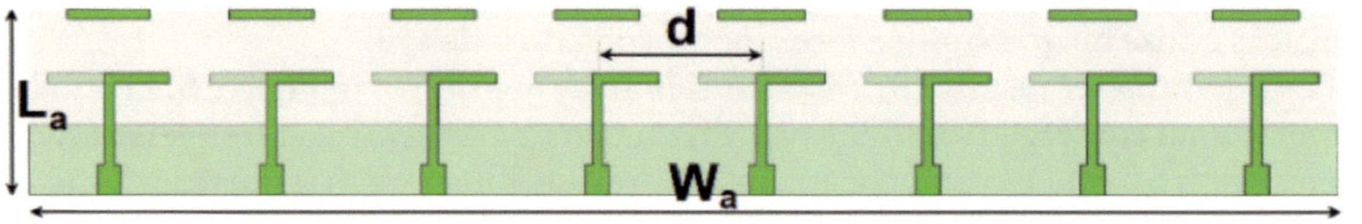

Figure 6.
Schematic of the designed THz phased array.

Moreover, the antenna resonators maintain high total efficiencies across their operational band, as evidenced in **Figure 7(b)**, which is crucial for maximizing signal strength and minimizing losses in high-frequency THz applications. These characteristics, wide bandwidth, low mutual coupling, and high efficiency, demonstrate the array's potential for point-to-point communications. **Figure 8** demonstrates the

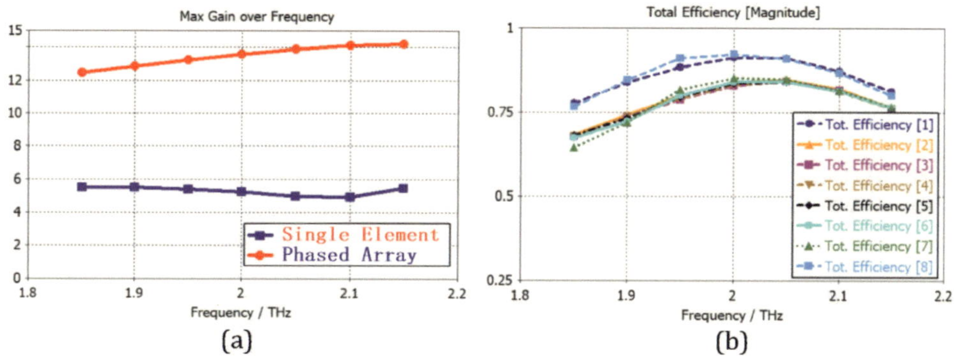

Figure 7.
(a) Gain level comparison and (b) total efficiency results.

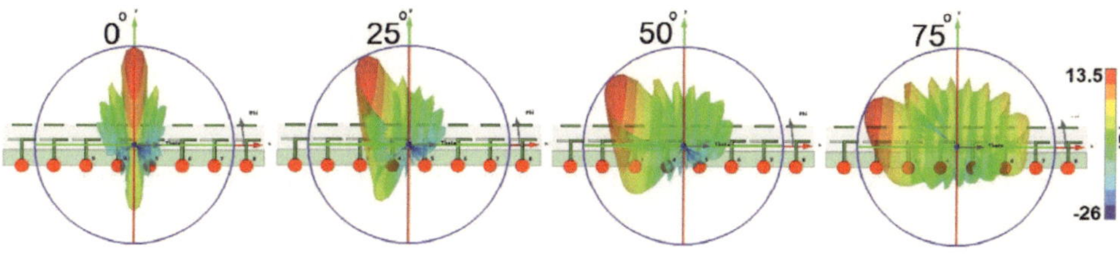

Figure 8.
3D beam-scanning at different scanning degrees (0 ~ 75).

beam-steering capabilities of the proposed array at the mid-frequency of 2 THz, showcasing a wide scanning range. This feature significantly enhances the array's versatility and adaptability for various communication scenarios [43]. Additionally, the antenna resonators exhibit high gain across multiple scanning angles from 0 to 60 degrees.

3. Conclusion

The review of THz antennas for future 5G and 6G networks highlights substantial advancements in design, materials, and integration techniques, addressing the unique challenges of terahertz frequencies. As the demand for ultra-high data rates, minimal latency, and improved spectral efficiency grows, THz antennas have emerged as key enablers, offering extensive bandwidth within the 0.1–10 THz range. Their diverse configurations—including single-port microstrip designs, 3D/non-planar structures, on-chip solutions, and multi-port arrays—demonstrate ongoing innovation tailored to next-generation wireless communication systems [44].

Microstrip antennas, valued for their compactness and integration potential, have been enhanced with advanced materials such as graphene and liquid crystalline polymers, reducing signal losses and improving efficiency. In parallel, 3D and non-planar structures, such as dielectric rod waveguides and corrugated horn antennas, provide high gain and directivity, making them ideal for THz imaging, sensing, and radar applications. On-chip antennas, leveraging CMOS and other silicon-based technologies, support miniaturized, high-performance, and cost-effective THz systems. Additionally, packaging techniques, including quad-flat no-lead (QFN) solutions and dielectric substrates, are improving performance and facilitating mass production. Multi-port and array THz antennas play a crucial role in high-capacity, high-speed 6G communication networks. These designs utilize array configurations and multi-port architectures to enhance gain, beam control, and polarization diversity, enabling applications such as wireless backhaul, radar, and advanced sensing. Phased-array antennas facilitate precise beam steering with high angular resolution, while frequency-dependent beam-scanning arrays offer high-gain solutions without complex active phase shifting. The continuous development of materials, micromachining techniques, and innovative feed structures is driving the next generation of THz antenna solutions. In addition to reviewing existing advancements, this chapter also explored the simulation and design of both single-element and array THz antennas, providing practical insights into their performance and implementation challenges. These simulations demonstrated key trade-offs between gain, bandwidth, efficiency, and integration, offering valuable perspectives on optimizing antenna designs for real-world applications. The design examples illustrated how different approaches—ranging from individual radiating elements to complex phased arrays—can be tailored to meet the stringent requirements of 6G systems, further emphasizing the critical role of THz antennas in future networks.

Despite these advancements, challenges remain, particularly in precision fabrication, material losses at THz frequencies, and seamless integration with existing communication systems. Designing compact, high-frequency antennas with efficient beam control and high radiation efficiency continues to be a focal area of research. Looking ahead, THz antennas will play a transformative role in shaping the future of 6G technology. With ongoing innovations in materials, fabrication techniques, and integration strategies, scalable and high-performance antenna solutions will become increasingly viable. These advancements will not only enable ultra-fast data transmission and low-latency communication but will also support emerging applications in high-resolution sensing, imaging, and next-generation radar systems. As research progresses, the focus will be on achieving seamless system integration, improved efficiency, and enhanced reliability, ensuring that THz antennas unlock the full potential of 6G networks across diverse real-world applications [45].

Author details

Haleh Jahanbakhsh Basherlou[1], Martins Odiamenhi[1], Naser Ojaroudi Parchin[1*], Chan Hwang See[1] and Ming Shen[2]

1 School of Computing, Engineering and the Built Environment, Edinburgh Napier University, Edinburgh, UK

2 Department of Electronic Systems, Aalborg University, Aalborg, Denmark

*Address all correspondence to: n.ojaroudiparchin@napier.ac.uk

References

[1] Jiang W et al. Terahertz communications and sensing for 6G and beyond: A comprehensive review. IEEE Communications Surveys & Tutorials. 2024;**26**(4):2326-2381

[2] Akyildiz IF, Kak A, Nie S. 6G and beyond: The future of wireless communications systems. IEEE Access. 2020;**8**:133995-134030

[3] Loscri V et al. The Road towards 6G: Opportunities, Challenges, and Applications. Switzerland: Springer Nature; 2024

[4] Jiang W, Schotten HD. Full-spectrum wireless communications for 6G and beyond: From microwave, millimeter-wave, terahertz to lightwave. In: 2023 IEEE 3rd International Conference on Computer Communication and Artificial Intelligence (CCAI), Taiyuan, China. 2023. pp. 353-357

[5] Mundlamuri R et al. 5G over terahertz using OpenAirInterface. In: 2024 19th Wireless on-Demand Network Systems and Services Conference (WONS), Chamonix, France. 2024. pp. 29-32

[6] Singh A et al. Wavefront engineering: Realizing efficient terahertz band communications in 6G and beyond. IEEE Wireless Communications. 2024;**31**(3):133-139

[7] Thomas S et al. A Survey on Advancements in THz Technology for 6G: Systems, Circuits, Antennas, and Experiments. arXiv preprint arXiv:2407.01957; 2024

[8] Tonouchi M. Cutting-edge terahertz technology. Nature Photonics. 2007;**1**:97-105

[9] Chan HA. Comparing wireless data network standards. In: AFRICON 2007. Windhoek, South Africa; 2007. pp. 1-15

[10] Nagatsuma T. Terahertz technologies: Present and future. IEICE Electronics Express. 2011;**8**(14):1127-1142

[11] He Y, Chen Y, Zhang L, Wong S-W, Chen ZN. An overview of terahertz antennas. China Communications. 2020;**17**(7):124-165

[12] Teng F, Wan J, Liu J. Review of terahertz antenna technology for science missions in space. IEEE Aerospace and Electronic Systems Magazine. 2023;**38**(2):16-32

[13] Pandya PR et al. Terahertz antennas—Review and design. In: Das S, Nella A, Patel SK, editors. Terahertz Devices, Circuits and Systems. Singapore: Springer; 2022

[14] Jamshed MA, Nauman A, Abbasi MAB, Kim SW. Antenna selection and designing for THz applications: Suitability and performance evaluation: A survey. IEEE Access. 2020;**8**:113246-113261

[15] Altaf A, Abbas W, Seo M. A wideband SIW-based slot antenna for D-band applications. IEEE Antennas and Wireless Propagation Letters. 2021;**20**(10):1868-1872

[16] Rabbani MS, Ghafouri-Shiraz H. Liquid crystalline polymer substrate-based THz microstrip antenna arrays for medical applications. IEEE Antennas and Wireless Propagation Letters. 2017;**16**:1533-1536

[17] Erdogan S et al. D-band integrated and miniaturized quasi-Yagi antenna

Array in glass interposer. IEEE Transactions on Terahertz Science and Technology. 2023;**13**(3):270-279

[18] Kushwaha RK, Karuppanan P. Investigation and design of microstrip patch antenna employed on PCs substrates in THz regime. Australian Journal of Electrical and Electronics Engineering. 2021;**18**(2):118-125

[19] Tajima T et al. 300-GHz step-profiled corrugated horn antennas integrated in LTCC. IEEE Transactions on Antennas and Propagation. 2014;**62**:5437-5444

[20] Generalov AA et al. Wide band mm- and sub-mm-wave dielectric rod waveguide antenna. IEEE Transactions on Terahertz Science and Technology. 2014;**4**(5):568-574

[21] Ranjkesh N et al. Millimeter-wave suspended silicon-on-glass tapered antenna with dual-mode operation. IEEE Transactions on Antennas and Propagation. 2015;**63**:5363-5371

[22] Yi H et al. 3-D printed millimeter-wave and terahertz lenses with fixed and frequency scanned beam. IEEE Transactions on Antennas and Propagation. 2016;**64**(2):442-449

[23] Alibakhshikenari M et al. A comprehensive survey on antennas on-chip based on metamaterial, metasurface, and substrate integrated waveguide principles for millimeter-waves and terahertz integrated circuits and systems. IEEE Access. 2022;**10**:3668-3692

[24] Ahmad WA et al. Planar highly efficient high-gain 165 GHz on-Chip antennas for integrated radar sensors. IEEE Antennas and Wireless Propagation Letters. 2019;**18**(11):2429-2433

[25] Kong S, Shum K-M, Chan CH. A 450-GHz wideband on-chip antenna with an extremely low profile. In: 2023 17th European Conference on Antennas and Propagation (EuCAP), Florence, Italy. 2023. pp. 1-3

[26] Bakshi HS et al. Low-cost packaging of 300 GHz integrated circuits with an on-chip patch antenna. IEEE Antennas and Wireless Propagation Letters. 2019;**18**(11):2444-2448

[27] Lee C, Jeong J. THz CMOS on-chip antenna Array using defected ground structure. Electronics. 2020;**9**:1137

[28] CST Microwave Studio, ver. 2022, CST, MA, USA, 2022

[29] Basherlou HJ. Dual-band MIMO antenna with monopole resonators for THz 6G communications. In: International Conference on Future Telecommunications and Artificial Intelligence (IC-FTAI). Egypt: Alexandria; 2024

[30] Deng R, Yang F, Xu S, Pirinoli P. Terahertz reflectarray antennas: An overview of the state-of-the-art technology. In: 2014 International Conference on Electromagnetics in Advanced Applications (ICEAA), Palm Beach, Aruba. 2014

[31] Yang Y, Gurbuz OD, Rebeiz GM. An eight-element 370-410-GHz phased-array transmitter in 45-nm CMOS SOI with peak EIRP of 8-8.5 dBm. IEEE Transactions on Microwave Theory and Techniques. 2016;**64**:4241-4249

[32] Kim H, Oh J. 140-GHz wideband array antenna-in-package using multimode resonance. IEEE Transactions on Antennas and Propagation. 2023;**71**(3):2136-2144

[33] Monroe NM et al. Electronic THz pencil beam forming and 2D steering

for high angular-resolution operation: A 98 × 98-unit 265GHz CMOS reflectarray with In-unit digital beam shaping and squint correction. In: 2022 IEEE ISSCC, San Francisco, CA, USA. 2022

[34] Guo K, Zhang Y, Reynaert P. A 0.53-THz subharmonic injection-locked phased array with 63- μ W radiated power in 40-nm CMOS. IEEE Journal of Solid-State Circuits. 2019;**54**(2):380-391

[35] Headland D et al. Integrated Luneburg and Maxwell fisheye lenses for the terahertz range. In: 2019 44th International Conference on Infrared, Millimeter, and Terahertz Waves (IRMMW-THz), Paris, France. 2019. pp. 1-2

[36] Li S et al. An eight-element 140 GHz wafer-scale phased-array transmitter with 32 dBm peak EIRP and > 16 Gbps 16QAM and 64QAM operation. In: 2021 IEEE MTT-S International Microwave Symposium (IMS), Atlanta, GA, USA. 2021

[37] Koyama Y et al. A high-power terahertz source over 10 mW at 0.45 THz using an active antenna Array with integrated patch antennas and resonant-tunneling diodes. IEEE Transactions on Terahertz Science and Technology. 2022;**12**:510-519

[38] Kim D et al. 64× 64-element and 32× 32-element slot array antennas using double-layer hollow-waveguide corporate-feed in the 120 GHz band. IEEE Transactions on Antennas and Propagation. 2014;**62**(3):1507-1512

[39] Jung D et al. Terahertz antenna-in-package design and measurement for 6G communications systems. IEEE Transactions on Antennas and Propagation. 2024;**72**(2):1085-1096

[40] Park CJ, Jung D, Seo J, Kwon TS, Ahn B, Wi S-H. Isolation improvement technique in dual-polarized antenna for sub-THz antenna-in-package. IEEE Antennas and Wireless Propagation Letters. 2023;**22**:3032-3036

[41] Sarabandi K, Jam A, Vahidpour M, East J. A novel frequency beam-steering antenna array for submillimeter-wave applications. IEEE Transactions on Terahertz Science and Technology. 2018;**8**:654-665

[42] Tekkouk K et al. Corporate-feed slotted waveguide array antenna in the 350-GHz band by silicon process. IEEE Transactions on Antennas and Propagation. 2017;**65**(1):217-225

[43] Basherlou HJ. A compact design of high-gain phased array antenna for THz and 6G communication networks. In: International Conference on Future Telecommunications and Artificial Intelligence (IC-FTAI). Alexandria, Egypt; 2024

[44] Mahmood A, Kiah MLM, Azizul ZH, Azzuhri SR. Analysis of terahertz (THz) frequency propagation and link design for federated learning in 6G wireless systems. IEEE Access. 2024;**12**:23782-23797

[45] Pant R, Malviya L. THz antennas design, developments, challenges, and applications: A review. International Journal of Communication Systems. 2023;**36**:1-39